Concepts of Modern Biology Series

William D. McElroy and Carl P. Swanson, Editors

Animal Parasitism, *Clark P. Read*
Behavioral Aspects of Ecology, *Peter H. Klopfer*
Bioenergetics, *Leonardo Peusner*
Concepts of Ecology, *Edward J. Kormondy*
Critical Variables in Differentiation, *Barbara E. Wright*
Euglenoid Flagellates, *Gordon F. Leedale*
Genetic Load: Its Biological and Conceptual Aspects, *Bruce Wallace*
An Introduction to the Structure of Biological Molecules,
J. M. Barry and E. M. Barry
Molecular Biology: An Introduction to Chemical Genetics,
J. M. Barry and E. M. Barry
The Organism as an Adaptive Control System, *John M. Reiner*
The Origins of Life on the Earth, *Stanley L. Miller and Leslie E. Orgel*
Processes of Organic Evolution, *G. Ledyard Stebbins*

Concepts of Modern Biology Series
William D. McElroy and Carl P. Swanson, Editors

The Origins of Life on the Earth

Stanley L. Miller

University of California, San Diego
La Jolla, California

Leslie E. Orgel

Salk Institute
La Jolla, California

Prentice-Hall, Inc., *Englewood Cliffs, New Jersey*

Miller, Stanley L
The origins of life on the earth.

(Concepts of modern biology series)
Includes bibliographical references.
1. Life—Origin. I. Orgel, Leslie E., joint author. II. Title. [DNLM: 1. Biochemistry. 2. Biogenesis. 3. Geology. QH325 M6510 1973]
QH325.M55 577 73-8651
ISBN 0-13-642082-6
ISBN 0-13-642074-5 (pbk.)

TO H. C. UREY

Current printing (last digit):

10 9 8 7 6 5 4 3 2 1

Printed in the United States of America

Prentice-Hall International, Inc., *London*
Prentice-Hall of Australia, Pty. Ltd., *Sydney*
Prentice-Hall of Canada, Ltd., *Toronto*
Prentice-Hall of India Private Limited, *New Delhi*
Prentice-Hall of Japan, Inc., *Tokyo*

Contents

Preface

The problem of the origin of life is truly interdisciplinary since an adequate discussion must present material that is normally found in courses in biology, chemistry, geology, and astronomy. Our aim in writing this book was to present as much of this material as possible in a form readily accessible to most science students at the junior and senior level and to some students at the freshman and sophomore level as well.

Readers who have not studied organic chemistry will certainly have difficulty with parts of Chapters 7, 8, 11, and 13. We see no way around this problem since the study of the origins of life is basically a study in organic synthesis. Only a book dealing with generalities can avoid the organic chemistry. Readers without a background in organic chemistry should still be able to learn about the state of the field by reading these chapters and ignoring those sections that deal with the details of the organic chemistry.

Chapter 4 and a few sections of other chapters involve considerable physical chemistry and elementary thermodynamics. Much of this material should be understandable by readers who have had a modern freshman chemistry course. Again, those readers who have difficulty with these sections should pass quickly over them.

The discussion of molecular biology in Chapters 6 and 12 is of necessity very much condensed. Biology students will probably be familiar with this material from their other courses; our discussion is written as a brief introduction for the chemistry and geology student, and is intended to be a review for the biology student.

We have not attempted to compile a complete or even a large list of references. We have only included what we consider the major papers in a given area along with a few supplementary citations. We believe that these will allow the interested reader to begin his literature search.

We are indebted to Mrs. Gwen Robbins and Mrs. Rosemary Brown for patient and skillful secretarial help and to Mr. Paul Forman for his assistance in preparing the figures. We are also indebted to Professors H. C. Urey, H. Craig, Y. Wolman, C. Sagan, R. Sanchez, and all who contributed helpful comments on all or parts of the manuscript.

STANLEY L. MILLER
LESLIE E. ORGEL

Chapter One

Introduction

It must be admitted from the beginning that we do not know how life began. It is generally believed that a variety of processes led to the formation of simple organic compounds on the primitive earth. These compounds combined together to give more and more complex structures until one was formed that could be called living.

No one should be satisfied with an explanation as general as this. We need a detailed theory that specifies the nature of the processes leading to the synthesis of organic compounds on the primitive earth, the nature of the compounds formed by these processes, and the quantities that could have accumulated abiogenically. Then we need to know the conditions under which the simplest organic compounds combined together to give monomers such as amino acids and nucleotides, and how these monomers condensed to polymers such as proteins and nucleic acids. At present, little of this detailed information is available.

It has been argued that the course of events leading to the appearance of the first living organism is essentially unknowable since no geological record of these events has been preserved. We do not accept this argument, for even if we concede the absence of any geological record, we still have experimental evidence of a kind. We are fairly certain that life did originate on earth, and we know in considerable detail the nature of the basic components and biosynthetic pathways that are common to all living organisms. While we cannot be certain that these compounds and mechanisms were important for the most primitive organisms, it is simplest to suppose that most of them were. Thus, information about the synthesis of important biochemicals, whether monomers or polymers, under primitive-earth conditions is likely to throw light on biochemical evolution.

It must be realized that our problem differs from those faced in most scientific work in that we are attempting to reconstruct a historical process. It is not possible to test a hypothesis concerning the origin of life by running rapidly through the entire process in the laboratory. We must therefore use different criteria in evaluating a theory. We ask that processes postulated in a theory of the origins of life be consistent with all accepted geological and astronomical data, and that each step be plausible in detail and be carried out in the laboratory insofar as possible. When a step cannot be investigated directly in the laboratory, say because it is too slow, related systems should be studied in such a way that extrapolation to primitive earth conditions is possible. This program is long and difficult. Quantitative equilibrium and kinetic data must be accumulated for many reactions and close attention must be paid to the geological evidence in order to define reasonable primitive earth conditions. When several prebiotic syntheses of the same compound are known, it is necessary to evaluate the relative importance of these different processes.

There may arise at some stage in this historical reconstruction the problem of deciding between two equally plausible but substantially or entirely different theories of the origins of life. How is a decision to be made in these circumstances? We can dispose of this problem by saying that we do not yet have one plausible, detailed, and complete hypothesis; we do not need to discuss the matter until we have two.

It has sometimes been suggested that life on earth was transported here from some other planet. According to the theory known as *Panspermia*, spores of extraterrestrial organisms were driven to the earth by the radiation pressure from another solar system. Recent calculations show that no viable organism could have survived the journey. It has also been claimed that no viable organisms could have arrived from outside the solar system in a meteorite. We do not find these arguments against Panspermia completely convincing, since we are not in a position to calculate reliably how often solid objects escape from a typical solar system. Therefore we should not be completely dogmatic in our belief that terrestrial life evolved from non-living matter on the earth, although all alternatives are much less plausible.

If life did not originate on earth, it may have evolved on some other planet where the conditions were very similar to those on the primitive earth, in which case the problem would be the same. On the other hand, the conditions may have been quite different. This would not dispose of the problem of the origins of life, but it would change it—we might well miss the most important prebiotic reactions if we restricted our attention to primitive-earth conditions. Despite these arguments, we believe that the chance that terrestrial life originated outside the earth is so small that we can ignore it. In our discussions of the origins of life, we shall consider only those reactions which could have taken place on the primitive earth.

We must mention at this point the relevance of the space program to studies of the origins of life. If there is an independent form of life on Mars, a comparison of Martian biochemistry with terrestrial biochemistry will provide a severe test of any theory of the origins of life. If, as may well be the case, we do not find another form of life in our solar system, we should nevertheless obtain valuable information concerning abiotic organic synthesis from studies of the chemistry of Mars and other planets.

In the first part of this book, we shall examine contemporary ideas on the formation of the earth and solar system in the light of relevant astronomical and geological data. We shall then discuss the primitive atmosphere and the ocean, and the synthesis of simple organic compounds and polymers on the primitive earth. This leads to a speculative discussion of the nature of the first living organisms and how they came into existence. Finally, we shall discuss the possibility that life exists elsewhere in the universe.

In the course of these discussions, we shall try to indicate whether or not we believe the data or theories to be firmly established. We shall be presenting our own ideas on the origin of life. Since there are many who hold different points of view, we cannot discuss all their ideas in detail. We shall outline some of the alternative theories, but in a short treatment of the field we cannot mention all of them.

General References to the Origin of Life

A. I. Oparin has published many books on the origin of life. The first appeared as early as 1924 (*Proischogdenie Zhizni*, Moscovsky Robotchii, Moscow), but was not widely available in the West. A translation of this pamphlet appears in J. D. Bernal, *The Origin of Life* (Weidenfeld and Nicolson, London; World Publishing Co., New York, 1967). Oparin's most important book is *The Origin of Life* (Macmillan, New York, 1938). This was reprinted by Dover in 1953 and called the 2nd edition. In this book, Oparin considers the origin of life beginning with a discussion of the primitive reducing atmosphere. He then discusses the synthesis of simple organic compounds, the formation of polymers, and the coacervate hypothesis of the first organism. This was a bold departure from the general thinking at the time, since most of the few people thinking about the subject felt the first organism was autotrophic and was able to synthesize all its cell material from CO_2 and H_2O. Oparin published further editions and revisions in 1957, 1962, 1964, and 1968. These subsequent editions are larger but not superior to the Macmillan and Dover editions.

Views similar to Oparin's but not as detailed or convincing were expressed independently by J. B. S. Haldane, *Rationalist Annual* **148**, 3 (1929); reprinted in *Science and Human Life* (Harper Bros., New York and London, 1933), p. 149.

R. Beutner, *Life's Beginning on the Earth* (Williams & Wilkins, Baltimore, 1938) appeared the same year as Oparin's first English edition and expressed a similar point of view. However, this book was not as widely read as Oparin's.

An important discussion of the heterotrophic origin of life proposed by Oparin was given by N. W. Horowitz, "On the evolution of biochemical synthesis," *Proc. Nat. Acad. Sci. U.S.* **31**, 153 (1945).

Important and influential extensions of Oparin's arguments were made by H. C. Urey, "The early chemical history of the earth and the origin of life," *Proc. Nat. Acad. Sci.* **38**, 351 (1952) [also in H. C. Urey, *The Planets* (Yale University Press, New Haven, 1952), pp. 149–157], and J. D. Bernal, *The Physical Basis of Life* (Routledge and Kegan Paul, London, 1951).

A widely read book which emphasized the thermodynamic aspects of life and its origin was H. F. Blum, *Time's Arrow and Evolution* (Princeton University Press, Princeton, 1951; 2nd Edition, 1955; 3rd Edition, 1968). Also very popular and frequently cited is G. Wald, "The Origin of Life," *Scientific American* **191**, 44–53 (August 1954). Four interesting articles by J. Haldane, J. D. Bernal, N. W. Pirie, and J. Pringle are contained in *New Biology* **16**, 12–67 (1954).

International symposia on the origin of life have been held in Moscow (1957), New York (1957), Walkulla Springs, Florida (1963), and Pont-à-Mousson, France (1970). The papers given at these symposia have been published: *The Origin of Life on the Earth*, A. I. Oparin, et al., eds. (Pergamon Press, New York, 1959); *Modern Ideas of Spontaneous Generation, Ann. New York Acad. Sci.* **69**, 225–376 (1957); *The Origin of Prebiological Systems*, S. W. Fox, ed. (Academic Press, New York, 1965); *Chemical Evolution and the Origin of Life*, R. Buvet and C. Ponnamperuma, eds. (North Holland, Amsterdam, 1971). The reader should realize that the quality and relevance of these papers are quite variable.

D. H. Kenyon and G. Steinman, *Biochemical Predestination* (McGraw-Hill, New York, 1969), is one of the best books on the subject, in spite of its title. Its level is somewhat higher than this book. R. M. Lemmon's review "Chemical evolution" in *Chemical Reviews* **70**, 95–109 (1970) is also excellent.

Biology and the Exploration of Mars, C. S. Pittendrigh, W. Vishniac and J. Pearman, eds. (*National Academy of Science Publication 1296*, Washington, D.C., 1966). This is a series of papers discussing topics such as the origin of life, the conditions on Mars, and the detection of life on Mars. A second volume, *Extraterrestrial Life—an Anthology and Bibliography*, compiled by E. A. Shneour and E. A. Ottesen (*National Academy of Science Publication 1296A*, Washington, D.C., 1966), reprints 34 papers and has a bibliography.

Other books include:

J. D. Bernal, *The Origin of Life* (Weidenfeld and Nicolson, London; World Publishing Co., New York, 1967).

M. Calvin, *Chemical Evolution* (Oxford University Press, New York, 1969).

J. Keosian, *The Origin of Life* (Reinhold, New York, 1964, 1st ed.; 1968, 2nd ed.).

M. G. Rutten, *The Origin of Life by Natural Causes* (Elsevier, Amsterdam, 1971). This book emphasizes the geological aspects.

A bibliography containing 1588 references has been compiled by M. W. West and C. Ponnamperuma, *Space Life Sci.* **2**, 225–295 (1970). A supplement to this bibliography with 212 references is in *Space Life Sci.* **3**, 293–304 (1972).

Chapter Two

The Formation of the Solar System

Any treatment of the conditions which existed on the primitive earth when life began must start by considering the way in which the solar system itself was formed. This procedure, incidentally, provides a natural basis for a discussion of the occurrence of life on other planetary systems (Chapter 14). Unfortunately, we know very little about many of the details in which we are most interested. The origin of the solar system is almost as hard to discuss as the origins of life. They are both historical processes of the type which cannot be repeated. While we can obtain some relevant evidence from laboratory experiments, the complete series of events that led to the formation of the solar system cannot be simulated.

Nevertheless, it is believed that the basic processes leading to the formation of the solar system are understood. Since the theory is complicated and a full discussion would be beyond the scope of our treatment, we shall consider only those general aspects of the problem which are most relevant to the origins of life.

It was formerly thought that the earth and other planets were formed when a star passed close to the sun. During this near collision, it was argued, material would have been torn from the sun and thrown into orbit around it. These fragments, after cooling, became the planets as we now know them. This theory of the formation of the solar system has been abandoned for three reasons. First, it is extremely unlikely (but not impossible) for two stars to have passed close enough, for they would have needed to approach to within a few star diameters of each other in order to form planet-sized bodies. Second, calculations are said to show that it is impossible to form planets by this means. Suitable pieces could have been ripped from the sun, but they would have travelled on hyperbolic paths instead of elliptical orbits and thus would have become objects free in space rather than parts

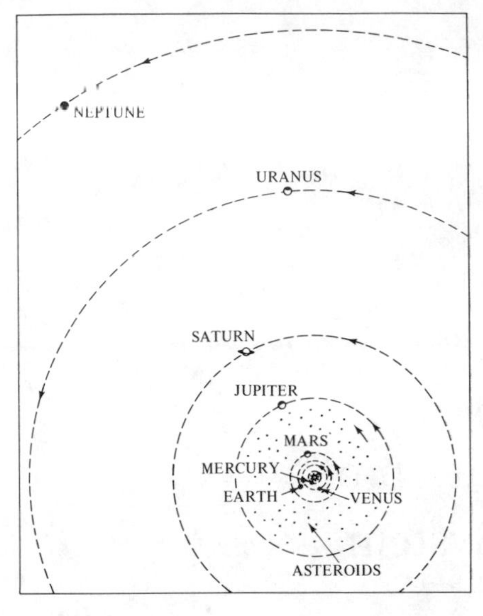

Fig. 2-1. The solar system.

of a solar system. Third, objects extracted from the sun in a near collision would have had so much kinetic energy that they would have disintegrated.

In spite of the total abandonment of the belief that the earth was formed in a molten state as a consequence of a near collision, we find that these ideas, particularly that of a completely molten earth, still persist in many popular writings and in some textbooks. The earth may have been molten at one time, but this is certainly not because it was torn from the sun.

All of the planets rotate in a single plane, which is the same as the plane of rotation of the sun. Most of the mass of the solar system is in the sun, but the planets carry most of the angular momentum. Any theory of the origin of the solar system must accommodate these facts and also the quantitative information obtained by astronomical observation. Table 2-1 summarizes some of the available data on the planets. The masses and radii are given in terms of the mass and radius of the earth. The distances are given in terms of the mean distance of the earth from the sun—the *astronomical unit* (AU).

The distances are known very accurately because they depend only on orbit determinations, which can be made precisely. The masses are known less accurately. The radii are sometimes difficult to measure since distortion by the atmosphere of the planets complicates observation of the planetary disks. Also, in the case of major planets, the atmosphere is very thick and an arbitrary decision must be made as to where the atmosphere ends and the planet begins. The densities of the planets are calculated from their masses and radii (density = mass/$\frac{4}{3}\pi r^3$), so uncertainties in the radii give corresponding uncertainties in the densities.

Table 2-1. Summary of Planetary Constants[a]

	Mean Distance from the Sun (astronomical units)[b]	Orbital Period (years)	Radius (Earth = 1)[c]	Rotation Period (d = days; h = hours)	Mass (Earth = 1)[e]	Average Density (g/cm^3)	Black Body Temperature (°K)[f]	Some Components of Atmosphere[g]	Major Components of Planet
Sun	–	–	109.1	25.38d	332,000	1.42		H, He	H, He
Mercury	0.39	0.24	0.382	58.4d	0.054	5.42	443	—	SiO_2, Fe
Venus	0.72	0.62	0.949	250d[d]	0.80	5.25	262	CO_2, (N_2), HCl, HF	SiO_2, Fe
Earth	1.00	1.00	1.000	23.93h	1.000	5.51	250	N_2, O_2, A	SiO_2, Fe
(Moon)	1.00	1.00	0.273	27.32d	0.0123	3.34	275	—	SiO_2, Fe
Mars	1.52	1.88	0.532	24.62h	0.105	3.96	218	CO_2, (N_2), H_2O	SiO_2, Fe
Jupiter	5.20	11.86	10.97	9.9h	310.2	1.33	105	H_2, (He), CH_4, NH_3	H_2, He
Saturn	9.54	29.46	9.03	10.3h	92.9	0.68	78	H_2, (He), CH_4, NH_3	H_2, He
Uranus	19.18	84.01	3.72	10.8h	14.2	1.60	55	H_2, (He), CH_4	H_2O, CH_4, NH_3
Neptune	30.07	164.8	3.50	15.8h	16.7	1.65	43	H_2, (He), CH_4	H_2O, CH_4, NH_3
Pluto	39.7	249.9	0.45?	6.4d(?)	0.1(?)	3.0(?)	42	?	(H_2O, CH_4, NH_3)?

[a] Data taken from F. L. Whipple, *Earth, Moon and Planets* (Harvard University Press. Cambridge, 1968). M. E. Ash, I. I. Shapiro, and W. B. Smith, *Science*, **174**, 551 (1971).

[b] The astronomical unit is 149.6×10^6 km = 92.9×10^6 miles.

[c] The mean radius of the earth is 6,378.2 km = 3,958.7 miles.

[d] The rotation of Venus is retrograde—i.e., in the opposite direction to the earth's rotation.

[e] The mass of the earth is 5.975×10^{27} g.

[f] The black body temperatures are the average for the planet rather than for the equator.

[g] Molecules expected to be present but not yet identified are in parentheses.

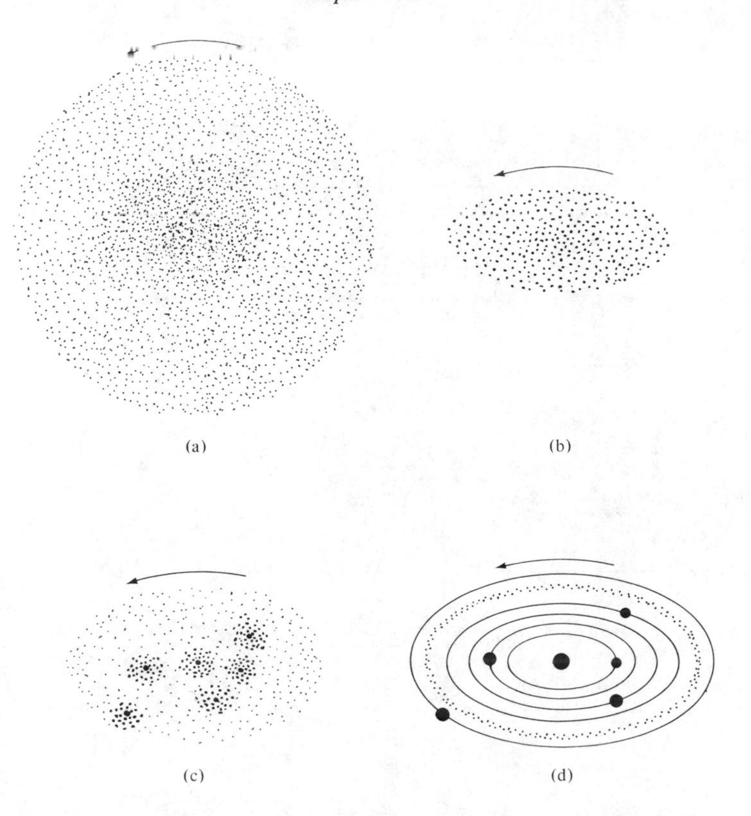

Fig. 2-2. Formation of solar system from cosmic dust cloud. (a) The spherical cloud has collapsed to a disk. (b) The material in the dust cloud is aggregating to form larger objects. (c) Aggregation is continuing and the objects are condensing to form the planets. (d) The aggregation is complete and the solar system has formed.

The value of the density ρ is an important clue to the chemical composition of a planet. It tells us, for example, that Jupiter and Saturn must be made up almost entirely of hydrogen and helium. The minor planets have much greater values of ρ than the major planets and therefore must be composed of more dense material, mostly iron, iron oxides, and silicon oxides.

The asteroids are a collection of small solid bodies which occupy orbits primarily between Mars and Jupiter. Their total mass is 0.04% that of the earth. It is thought by some that the asteroids were formed by the collision of two small planets, but there are alternative theories of their origin. One such theory, for example, is that they represent an accumulation of small objects which have been prevented from condensing into a single large planet.

The current theory of the origin of the solar system is a modern version of the Kant-Laplace theory (Figure 2-2). It envisions the formation of the planetary system as a natural consequence of the formation of the sun itself. It is assumed that the material which formed the solar system was at one time widely dispersed as a diffuse cloud of cosmic dust. This cloud is assumed to have had some angular momentum. The total angular momentum of all the matter originally in the dust cloud must have been conserved during the subsequent contraction. Since the angular momentum of the sun is now small, most of the angular momentum of the cloud must have been transferred to the planets and to the hydrogen and helium that escaped from the solar system.

Calculations show that such a cloud of cosmic dust would have collapsed into a disk when its density and angular velocity reached certain critical values. The orientation of this disk would have determined a unique plane of rotation for the planets and the sun in a very natural way. According to this theory the central portion of the disk or nebula must have had a greater density than the outer part. Such a situation would be unstable since the gravitational attraction of the dense core would cause further accretion of matter to it.

What prevented the whole nebula from condensing to a single central core, the sun? As the density of the core increased, gravitational energy was released, causing a substantial amount of energy to be released as radiation. The pressure exerted by this radiation acted on gaseous molecules and dust particles in the nebula forcing them outwards and preventing a complete collapse. During the later stages of the contraction of the core, radiation would have been produced by nuclear processes.

The final condensation of material in concentric bands around the sun into the planets is poorly understood. We do not know how the small particles of dust aggregated to form large masses. The process by which the angular momentum, which was initially concentrated in the central region of the nebula, was transferred to the planets is also uncertain. It may have involved the interaction of charged particles with the magnetic field of the nebula.

This theory provides a partial explanation of the masses and densities of the planets. The inner planets are small and composed largely of non-volatiles because the temperatures were high in the central region of the dust cloud at the time the solar system was formed. Volatile compounds such as CH_4, NH_3, H_2O, H_2, and He escaped from this region or were blown away by the radiation pressure from the sun. It is interesting that the elements most depleted on the inner planets are just those which form volatile compounds under reducing conditions. A more detailed discussion of the earth's atmosphere follows in the next section.

In the region now occupied by Jupiter and Saturn, the temperatures and radiation pressure were lower. This allowed the accumulation of massive bodies containing large amounts of hydrogen and helium. Because of the low temperatures and large gravitational fields, these gases could not escape after the planets had accumulated. Consequently, the abundances of elements on the major planets are approximately the same as their cosmic abundances.

The densities of Uranus and Neptune suggest that they are made up largely of CH_4, NH_3, and H_2O, and that they contain much less hydrogen and helium than the Jovian planets. We do not understand why hydrogen and helium were lost from this region of the solar system, since the temperature was even lower than in the neighborhood of Jupiter and Saturn. Perhaps the gas pressures were lower at the limits of the dust cloud.

We are particularly interested in the temperature in the region of the planets during their condensation. In the case of the earth, this temperature is believed by some workers to have been less than 300°C on the average. Higher temperatures would have removed nearly all the volatile compounds, such as H_2O, nitrogen compounds and rare gases, from the dust particles, while in fact they were in part retained. These considerations imply that the earth was formed at quite a low temperature. There are some workers who disagree with this conclusion and believe that the cosmic material in the region of the earth was much hotter before and during the condensation of material into the earth.

The formation of the sun and planets is thought to have occurred rather quickly over a period of 60 to 260 million years. This is deduced from the xenon content of meteorites which are believed to be about as old as the solar system. The half-life for the production of Xe^{129} from the decay of I^{129} is about 16 million years. Since the meteorites contain an excess of Xe^{129} which must have arisen from I^{129}, we infer that they must have solidified within a few half-lives of the formation of I^{129}. Thus the meteorites, and hence the planets, were formed during or shortly after the synthesis of some of the heavy elements that make up the solar system.

THE ORIGIN OF THE EARTH'S ATMOSPHERE

We have already seen that volatile elements are severely depleted on the earth. A great deal can be learned about the early history of the atmosphere from quantitative data. The cosmic abundances of the rare gases, carbon, nitrogen, and oxygen, as well as the amounts of these elements on the earth, are given in Table 2-2, together with data for some nonvolatile elements. Only A^{36} and A^{38} are included, since the large amount of A^{40} in the atmosphere comes from the decay of K^{40}. Also included in the table are the depletions relative to xenon. These are defined as the ratios of the abundances of elements on the earth to their cosmic abundances, divided by the corresponding ratio for xenon (4×10^{-5}). The last column in the table gives an estimate for each element of the fraction of material which was originally in the dust cloud that is retained on the earth.

It is immediately clear that the volatile elements have escaped to very different extents. This fractionation must have occurred before the earth condensed into a solid body of about its present size, since we know that, apart from hydrogen and helium, these gases cannot escape from the earth's gravitational field.

Table 2-2. Depletion Factors of Volatile Elements on the Earth Relative to Cosmic Abundances

	Cosmic Abundance (Si = 10,000)	mole cm^{-2} on the Earth	Depletion Relative to Xenon	Fraction Retained on the Earth
Ne	24,000	6.5×10^{-4}	6.8×10^{-4}	6.8×10^{-11}
$A^{36} + A^{38}$	1,000	1.2×10^{-3}	3.0×10^{-2}	3.0×10^{-9}
Kr	0.55	3.5×10^{-5}	1.6	1.6×10^{-7}
Xe	0.07	2.8×10^{-6}	1.0	1.0×10^{-7}
H_2O	236,000	15,000	1,600	1.6×10^{-4}
C	137,000	350	64	6.4×10^{-6}
N	24,000	54	56	5.6×10^{-6}

The easiest way to account for the great depletion of neon relative to xenon is to suppose that neon escaped from a relatively weak gravitational field, but that xenon, which has a much higher atomic weight, was retained. However, if neon could escape from this early gravitational field, so could CH_4, NH_3, and H_2O, since their molecular weights are similar to the atomic weight of neon. To account for the amounts of carbon, nitrogen, and oxygen that were retained, we must assume that they were held mainly in a chemically combined form: Nitrogen was held in ammonium silicates, other ammonium compounds, and perhaps nitrides; carbon could have been present as graphite, carbides, and perhaps carbonates; water could have been present in hydrated minerals.

The molecules CN, C_2, and CH have been identified spectroscopically in the tails of comets. A number of organic molecules have been identified in cosmic dust clouds as well (Chapter 15). It is likely, therefore, that organic compounds were synthesized in the cosmic dust cloud from CH_4, NH_3, and H_2O and contributed to the retained carbon and nitrogen. We cannot estimate how much of the CH_4 in the dust cloud was converted into organic compounds before the earth was formed, but it could have accounted for a large fraction of the carbon retained on the earth. Some nitrogen could have been held the same way. Organic materials in the carbonaceous chondrites provide a plausible model for these organic compounds (see Chapter 15).

There are many theories that attempt to describe the events that took place during the condensation of cosmic dust and larger objects into the earth. These theories differ greatly from each other and tend to be rather detailed. At one extreme, it is believed that the accumulation of material had to have been very rapid so that gravitational energy was released rapidly. The gravitational energy released during the accretion of material into the earth would have been sufficient to melt the entire earth, including the surface, unless this energy was radiated away faster than it was released by the incoming material. At the other extreme are theories in which the accretion of material is supposed to have been slow, so that the energy released could have been dissipated by radiation at a rate comparable to the rate of energy production by incoming material. In this model the entire earth would not have melted. However, the interior would have melted because of adiabatic compression as more material accumulated and also because of the heat generated by decay of radioactive elements. Thus, in both models a molten core would have formed.

At the present time, opinion favors models that predict an entirely molten earth; but opinion has vacillated in this area: ten years ago the concept of a nonmolten earth was generally in favor, whereas ten years before that the opposite conclusion was popular.

Whether the entire earth was molten or not does not greatly affect our discussions of the prebiotic synthesis of organic compounds. If the entire earth melted, all the organic compounds, both in the interior and on the surface, would have been pyrolyzed completely to an equilibrium mixture of CO, CH_4, H_2, N_2, NH_3, and H_2O. When the earth cooled down sufficiently, the crust would have formed. When the average temperature on the surface and in the atmosphere was low enough, organic compounds would have been synthesized and, most important, would have accumulated.

According to models that do not involve a completely molten earth, the sequence of events would have been different. The carbon, both inorganic and organic, would have been distributed more or less evenly in the interior of the earth, since the infalling material would contain an approximately constant percentage of carbon. Much of the carbon and nitrogen must have escaped from the interior, since most of the earth's nitrogen, carbon, and water is believed to be concentrated at the surface. It is clear from the presence of radiogenic argon in the atmosphere that much, but not necessarily all, of the earth's interior has degassed. The most plausible mechanism for this involves volatilization. The interior of the earth was undoubtedly very hot and probably molten. Thus volatile material such as the noble gases, N_2 and CO, could have been driven to the surface. The organic compounds would have been pyrolyzed and the decomposition products driven to the surface.

Most organic compounds of biological interest such as amino acids, purines, and pyrimidines are nonvolatile and so would have been destroyed in the earth's interior. Eventually, a mixture of CH_4, CO_2, CO, NH_3, H_2O, N_2, and H_2 would have escaped to the surface and become part of the primitive atmosphere.

We do not know much about the composition of the mixture which reached the surface, since equilibrium in the C, H, O and N, H, O systems is strongly temperature dependent. It is likely that outgassing took place at temperatures between 300° and 1,500°C. At the higher temperatures, CO and N_2 are stable, but at the lower temperatures the more stable species are CH_4 and NH_3. In either case, we believe that soon after entering the atmosphere the outgassed material equilibrated, or more correctly, adjusted

Fig. 2-3. A newly emerged volcano erupting off the coast of Iceland in 1964. This photograph gives some impression of the vast quantities of gas evolved from a typical volcano. (Courtesy James Crowe, NAVOCEANO. From Stokes, *Essentials of Earth History*, Prentice-Hall, Inc., Englewood Cliffs, New Jersey, 1966.)

to a steady state, dictated by the temperature, pressure, and radiation density in the atmosphere.

Volcanoes are assumed to have been responsible for most of the outgassing of the earth. It is often said that the amount of volcanic activity on the primitive earth was very much greater than it is at present. However, while this may be true, there is no geological evidence for this. We do not know how long the degassing of the interior of the earth took. Estimates vary from 100,000 to several billion years.

Gases evolving from present-day volcanoes are quite variable in their composition. Some contain mainly CO_2 and SO_2 and are oxidizing, while others contain substantial amounts of CO, H_2, and CH_4 in addition to CO_2 and are reducing. Most contemporary volcanic gases are produced from melted crustal rocks and recirculated ground water. It has not been proved that any of the gases which are found in volcanoes come from the interior of the earth. Thus, the composition of present-day volcanic gases tells us little about the formation of the primitive atmosphere.

In addition to the material degassed from the interior, part of the primitive atmosphere may have been formed during an earlier period when massive objects were colliding with the partly formed earth. The heat generated in

Table 2-3. Selected Examples of Volcanic Gases. The figures for the gases—excluding water—are in volume percent. The value for water is its percentage of the total gases. Since air is usually present in the samples collected, the figures in the table have been corrected for this air.[a]

Volcanic gases are extremely variable in their composition, even from different fumaroles in the same area or from the same fumarole at different times. All the evidence indicates that no volcano on the continents is emitting gases from the mantle (referred to as juvenile gases): The gases seem to be produced from the decomposition of material in crustal rocks or from ground waters that have seeped down and have then been heated. Juvenile gases may be present in volcanoes on the ocean floor, but no samples from such volcanoes have been obtained.

Location	CO_2	CO	CH_4	NH_3	H_2	HCl	HF	H_2S	SO_2	H_2O
White Island, New Zealand (Big Donald Fumarole)	57.9	–	0.5	–	41.5	–	–	–	–	–
Nyerogongo lava lake, Congo	84.4	5.1	–	–	1.6	–	–	–	9.0	43.2
Mount Hekla, Iceland (shoulder crater)	23	3	–	–	16	52	trace	–	6	–
Lipari Island, Italy, Trachyte Volcano Fumarole	93.0	–	–	–	–	0.5	–	2.9	3.6	98.9
Larderello, Italy	92.7	–	0.92	1.72	1.76	–	–	2.45	–	–
Zavaritskii Crater, Kamchatka, USSR (crater wall)	–	67	–	–	–	33	–	–	–	–
Same, Crater B-1	–	21	–	–	42	25	–	–	12	–
Unimak Island, Alaska, Okmok Crater	47	–	–	–	–	–	–	–	53	95

[a] Taken from *Data of Geochemistry, Volcanic Emanations*, Geological Survey Professional Paper 440-K (U.S. Government Printing Office, Washington, D.C., 1963).

the collisions would have led to extensive degassing, particularly of the incoming material.

In summary, much of the carbon, nitrogen, and water on the earth were retained as nonvolatile compounds when the earth was formed from the cosmic dust cloud. The carbon was most likely present in part as organic compounds, but these were homogeneously distributed throughout the interior of the earth. When the carbon, nitrogen, and water were brought to the surface by degassing, most organic compounds were pyrolyzed. The gases mostly came to the surface from volcanoes, but we do not know how long it took, nor do we know much about the composition of volcanic gases on the primitive earth.

References

There are many books and review articles that deal in whole or in part with the origin of the solar system and the earth. A few of these are:

H. C. Urey, *The Planets: Their Origin and Development* (Yale University Press, New Haven, 1952).

H. C. Urey, "Chemical evidence relative to the origin of the solar system," *Monthly Notices Roy. Astron. Soc.*, London, **131**, 199 (1966).

R. Jastrow and A. G. W. Cameron, eds., *Origin of the Solar System* (Academic Press, New York, 1963).

H. E. Suess, "Chemical evidence bearing on the origin of the solar system," *Ann. Review of Astronomy and Astrophysics*, **3**, 217–234 (1965).

P. J. Brancazio and A. G. W. Cameron, eds., *The Origin and Evolution of Atmospheres and Oceans* (John Wiley & Sons, New York, 1964).

H. Alfvén and G. Arrhenius, "Origin and evolution of the solar system I and II," *Astrophysics Space Sci.* **8**, 338–421 (1970); **9**, 3–33 (1970).

A detailed description of the solar system can be found in:

G. P. Kuiper, ed., *The Solar System*, 4 volumes (University of Chicago Press, Chicago, 1953–1963).

Discussions of the accumulation material to form the earth and its core include:

D. L. Anderson, C. Sammis, and T. Jordan, "Composition and evolution of the mantle and core," *Science* **171**, 1103 (1971).

P. J. Hart, ed., *The Earth's Crust and Upper Mantle* (American Geophysical Union, Washington, D.C., 1969).

The fractionation of the noble gases between the solar nebula and the earth and its importance in consideration of the origin of the earth was pointed out in:

H. Suess, *J. Geology* **57**, 600 (1949).

H. Brown, in *The Atmospheres of the Earth and Planets*, G. P. Kuiper, ed. (University of Chicago Press, Chicago, 1949 and 1952), pp. 258–266.

Table 2-2 is based on these papers, but more recent estimates of the atomic abundances have been used (H. C. Urey, *Ann. N.Y. Acad. Sci.* **194**, 75 (1972).

Chapter Three

Geological Evidence

We could, in principle, test the correctness of our attempts to reconstruct the steps leading to the origin of life, if we had sufficient geological evidence. As we shall show, this evidence is not available, since no sedimentary rocks are known that were laid down during the period when life began. In spite of this, the available geological record has important things to tell us.

THE GEOLOGICAL TIME SCALE

The most important information that the geological record gives us is the time scale for the origin of life. This time scale is derived by dating rocks and meteorites. Until recently it had only been possible to give relative ages for series of rocks—by comparing the fossils preserved in them, for example. Another method depended on the simple principle that sediments are deposited in order, the lowest one being the oldest. Care has to be taken in applying this principle however, since sediments are frequently turned on edge, and in some cases overturned, in the course of mountain building. However, geologists have been able to unravel such sequences of events by making sufficiently detailed observations of rock series.

Knowledge of the relative ages of rocks is very useful, but we are more interested in their absolute ages. From our point of view, the most important advance in geology in the last 20 years is the development of techniques which enable us to determine the absolute ages of rocks.

There are three principle methods for obtaining this information. One method is potassium-argon dating which depends on the decay of the naturally occurring K^{40} isotope to A^{40} and Ca^{40}.

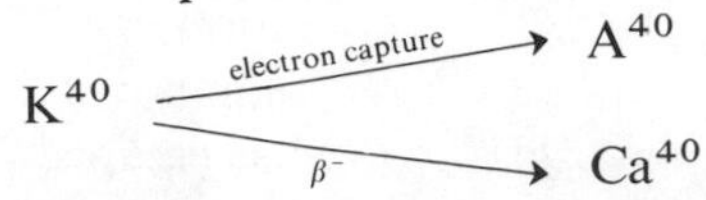

Fig. 3-1. The Grand Canyon from the Colorado River. In the foreground are Precambrian rocks that are approximately 2 billion years old. In the distance are the Tapeats sandstone (Cambrian), Muav limestone (Cambrian), and the Kaibob limestone (Permian).

The half-life ($t_{1/2}$) of K^{40} is 1.26×10^9 years. About 90% of the K^{40} decays to Ca^{40} and the remaining 10% decays to A^{40}. Since the age of the earth is longer than this, not much K^{40} survives in the older sediments, but there is enough to make the dating method work. The age of the rock is obtained by measuring both the A^{40} content of the rock and the potassium content. Provided the rock is assumed to have contained no argon when it was laid down, the age of the rock may be shown to be

$$t = \frac{1}{\lambda} \ln \left[1 + \frac{A^{40}}{K^{40}} \left(\frac{1 + R}{R} \right) \right]$$

where R is the ratio (0.11) of A^{40} to Ca^{40} produced in the decay of K^{40}, and λ is the decay constant ($0.693/t_{1/2}$) in the formula $N = N_0 e^{-\lambda t}$ which relates the number of surviving atoms (N) to the number of atoms originally present (N_0). This estimate is not valid if argon has escaped.

THE GEOLOGIC TIME SCALE

ERA	PERIOD	EPOCH	MILLIONS OF YEARS AGO (APPROX.)	DURATION IN MILLIONS OF YEARS (APPROX.)
CENOZOIC	QUATERNARY	RECENT PLEISTOCENE	0-1	1
	TERTIARY	PLIOCENE	1-13	13
		MIOCENE	13-25	12
		OLIGOCENE	25-36	11
		EOCENE	36-58	6
		PALEOCENE	58-63	
MESOZOIC	CRETACEOUS		63-135	72
	JURASSIC		135-181	46
	TRIASSIC		181-230	49
PALEOZOIC	PERMIAN		230-280	50
	PENNSYLVANIAN		280-310	30
	MISSISSIPPIAN		310-345	35
	DEVONIAN		345-405	60
	SILURIAN		405-425	20
	ORDOVICIAN		425-500	75
	CAMBRIAN		500-600	100
PRECAMBRIAN	UPPER MIDDLE LOWER	Although many local subdivisions are recognized, no world-wide system has been evolved. The Precambrian lasted for at least 2½ billion years. Oldest dated rocks are at least 2,700 million, possibly 3,300 million, years old.		

RELATIVE DURATIONS OF MAJOR GEOLOGICAL INTERVALS
CENOZOIC
MESOZOIC
PALEOZOIC
PRECAMBRIAN

Figure 3-2

It is technically difficult to measure the small amount of argon in a rock, but the procedure has been worked out and is routinely used in many geology laboratories. The possibility that argon escaped from the rock after it was laid down is more difficult to eliminate since argon can diffuse through and thus escape from many potassium minerals, especially clays and micas. For this reason, many rocks cannot be dated by the potassium-argon method.

A second method of dating is the rubidium-strontium method. This is based on the decay of naturally occurring Rb^{87} by the emission of an electron (β^-). The half-life is 4.7×10^{10} years.

$$Rb^{87} \xrightarrow{\beta^-} Sr^{87}$$

The age of the sample is given by

$$\text{Age} = \frac{\%\text{ radiogenic } Sr^{87}}{\%\ Rb^{87}} \frac{t_{1/2}}{0.693}$$

The amount of Rb^{87} and Sr^{87} is measured with a mass spectrometer. Although there is always some nonradiogenic strontium in a rock, if the amount is small it can be estimated from the amount of nonradioactive Sr^{86}. This method requires the use of minerals that are rich in rubidium and poor in strontium. There are, of course, complications with this dating technique. The rock must not have been heated for long periods because Rb and Sr could then have diffused out of the mineral and we would derive a false age. The rock also must not have been leached with water.

The third method of dating is the uranium-thorium-lead method, which depends on the following decay scheme.

$$U^{238} \xrightarrow[t_{1/2} = 4.51 \times 10^9 \text{ yr}]{\alpha} Th^{234} \rightarrow \begin{matrix}\text{17 further}\\ \text{radioactive steps}\end{matrix} \rightarrow Pb^{206}$$

$$U^{235} \xrightarrow[t_{1/2} = 7.1 \times 10^8 \text{ yr}]{\alpha} Th^{231} \rightarrow \begin{matrix}\text{15 further}\\ \text{radioactive steps}\end{matrix} \rightarrow Pb^{207}$$

$$Th^{232} \xrightarrow[t_{1/2} = 1.39 \times 10^{10} \text{ yr}]{\alpha} Ra^{228} \rightarrow \begin{matrix}\text{10 further}\\ \text{radioactive steps}\end{matrix} \rightarrow Pb^{208}$$

The uranium and thorium in the sample are determined by chemical analysis; the isotopic composition of the lead is determined with a mass spectrometer. Pb^{204}, a stable isotope which is not formed in any known radioactive decay, is always present and is used for comparing isotope ratios.

The age of the rock is determined by considering the growth of lead due to uranium decay

$$(Pb^{206})_{\text{present}} = (Pb^{206})_{\text{original}} + U^{238}_{\text{present}}(e^{\lambda_{238}t} - 1)$$

where $\lambda_{238} = 0.693/t_{1/2}$ is the decay constant for the α-decay of U^{238}, and the Pb and U refer to the number of atoms of each isotope. The time is measured from when the mineral was laid down or became a closed system. Isotopic concentrations are usually measured relative to the concentration of Pb^{204}, which is constant, so we divide through by the nonradiogenic Pb^{204} to give

$$\left(\frac{Pb^{206}}{Pb^{204}}\right)_{\text{measured}} = \left(\frac{Pb^{206}}{Pb^{204}}\right)_{\text{original}} + \left(\frac{U^{238}}{Pb^{204}}\right)_{\text{present}} (e^{\lambda_{238}t} - 1)$$

A similar equation for Pb^{207} may be written:

$$\left(\frac{Pb^{207}}{Pb^{204}}\right)_{\text{measured}} = \left(\frac{Pb^{207}}{Pb^{204}}\right)_{\text{original}} + \left(\frac{U^{235}}{Pb^{204}}\right)_{\text{present}} (e^{\lambda_{235}t} - 1)$$

These two equations permit the calculation of the age of the sample as long as little lead was present in the rock or mineral when it was laid down.

In practice, things are quite often more complicated, so that age calculations involve elaborate plotting of the experimental data. This dating method always gives two independent ages, and occasionally a third if the Pb^{208} arising from the decay of thorium is considered. If the two uranium ages agree, the age determination is considered reliable.

Each of these dating methods has its limitations. The potassium-argon method is the most general since potassium occurs in almost all rocks. But since many potassium minerals leak out their radiogenic argon, marine sediments usually cannot be dated by this method. The rubidium-strontium method and the uranium-lead method are frequently not applicable to a particular rock because there are not suitable minerals in it. This applies especially to limestones.

It is frequently possible to date a group of rocks by a combination of all three methods, even though a single rock cannot usually be studied in this way. A number of rocks are taken from a region so small that they are sure to have the same age. The three methods are then applied as appropriate. If all three dating methods give the same age, the age is taken to be quite reliable for the association of rocks.

In the case of such rocks as limestone, which cannot be dated directly, some information about age can often still be obtained. It frequently happens that a limestone series has been cut by a flow of lava or other igneous rock which can be dated. Since the limestone must have existed before it was cut by the igneous rock, the limestone must then be older than the igneous rock. If the limestone cuts the igneous rock, the limestone must have been laid down after the formation of the igneous rock, and the igneous rock must be older. This is a qualitative method which tells only the relative age of two rocks. Therefore, it is frequently possible to state only that a limestone is, for example, older than 2.7×10^9 years, since no estimate can be given as to whether it is one million or one billion years older. This incomplete information is better than none at all.

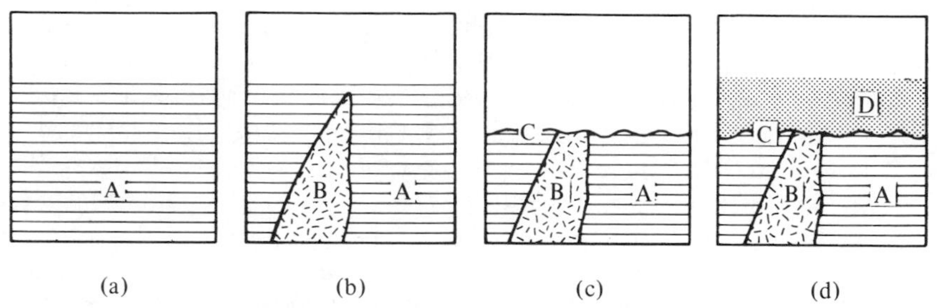

Fig. 3-3. Absolute and relative dating. The blocks illustrate a succession of geologic events that present a typical problem in dating. Sediments A are deposited in regular order. Later, intrusion B, judged to be 20 million years old according to the radioactive materials it contains, is intruded. A lengthy period of erosion produces the unconformity C, which cuts away part of the intrusion. Finally, the second set of sediments, D, is laid down. By superposition, we know that the relative ages are: A, B, C, D. The absolute date given by the intrusion indicates that beds A are older than 20 million years and that unconformity C and beds D are younger than 20 million years. Obviously, we can date the sediments more precisely if they contain minerals suitable for radiometric measurements. (From Stokes, *Essentials of Earth History*, Prentice-Hall, Inc., Englewood Cliffs, New Jersey, 1966.)

THE AGE OF THE EARTH

The age of the earth is usually taken to be the same as the age of the meteorites. This assumption is based on the generally held view that the sun, planets, and meteorites were formed from the same cloud of cosmic dust at about the same time. The error in our estimate of the age of the earth would be quite unimportant in the following discussion if the meteorites were formed one million or even 100 million years after the formation of the earth.

All three dating methods have been used to estimate the age of meteorites. The stony meteorites have been studied more often than the irons since the latter contain very little potassium, rubidium, and uranium. The three dating methods all give the same age, 4.6×10^9 years, which makes this age seem very reliable.

The age of the earth can be obtained independently by the uranium-lead dating method. This procedure makes a number of assumptions about the way the earth differentiated after it had formed. By analyzing rocks of different ages for their lead isotope content, it is possible by a complicated procedure to estimate the age of the earth. The value obtained is close to the 4.6×10^9 years determined for the age of the meteorites.

THE AGE OF THE MOON

It has always been thought that the moon was formed at about the same time as the earth (except in theories where the moon was formed by the breaking up of the earth). The results of dating lunar samples generally confirm this picture.

The preliminary reports of the ages of the first lunar rocks (Apollo 11) gave values of 2.5 to 3.5×10^9 years by the potassium-argon method. It was shown later that these ages were low because of loss of argon. The currently accepted age of the Apollo 11 rocks is 3.6×10^9 years; this is the time that the material was last remelted. The ages of the rocks from the Apollo 12, 14, and 15 missions are 4.5×10^9 years; the lunar soils from these missions are dated as 4.6×10^9 years. These dates are primarily rubidium-strontium ages with a few uranium-lead ages.

We therefore have three independent ages for the formation of the earth and solar system—the meteorites, the lead isotopes in the crust of the earth, and rocks from the surface of the moon. The age of the solar system and earth, therefore, seems well established as 4.6×10^9 years. Although the accepted values may change by a hundred million years or so as the dating methods are refined, such changes would have no effect on considerations of the origin of life.

Fig. 3-4. A well-preserved trilobite, *Elrathia*, from the Middle Cambrian of Utah. Many thousands of perfect specimens of this species, ranging in size from $\frac{1}{8}$ in. to 2 in. long, have been collected at a site in the House Range, western Utah. (From Stokes, *Essentials of Earth History*, Prentice-Hall, Inc., Englewood Cliffs, New Jersey, 1966.)

THE GEOLOGICAL PERIODS

The various geological periods were originally ordered on the basis of their fossils, but they have now been given absolute ages, as shown in Fig. 3-2. These determinations of absolute ages have confirmed the earlier ordering. There are still some uncertainties in a few of the ages, but any errors are small and unimportant for our purposes.

The first period that we shall consider is the beginning of the Cambrian, 600 million years ago. This is referred to as the base of the Cambrian because its sediments are usually found at the bottom of sequences of sediments laid down in the Cambrian period. The fossils of hard-shelled organisms survive from the Cambrian, but not from earlier periods in geological time. Until recently, the Cambrian was the earliest period for which there was clear evidence of the presence of any living organisms.

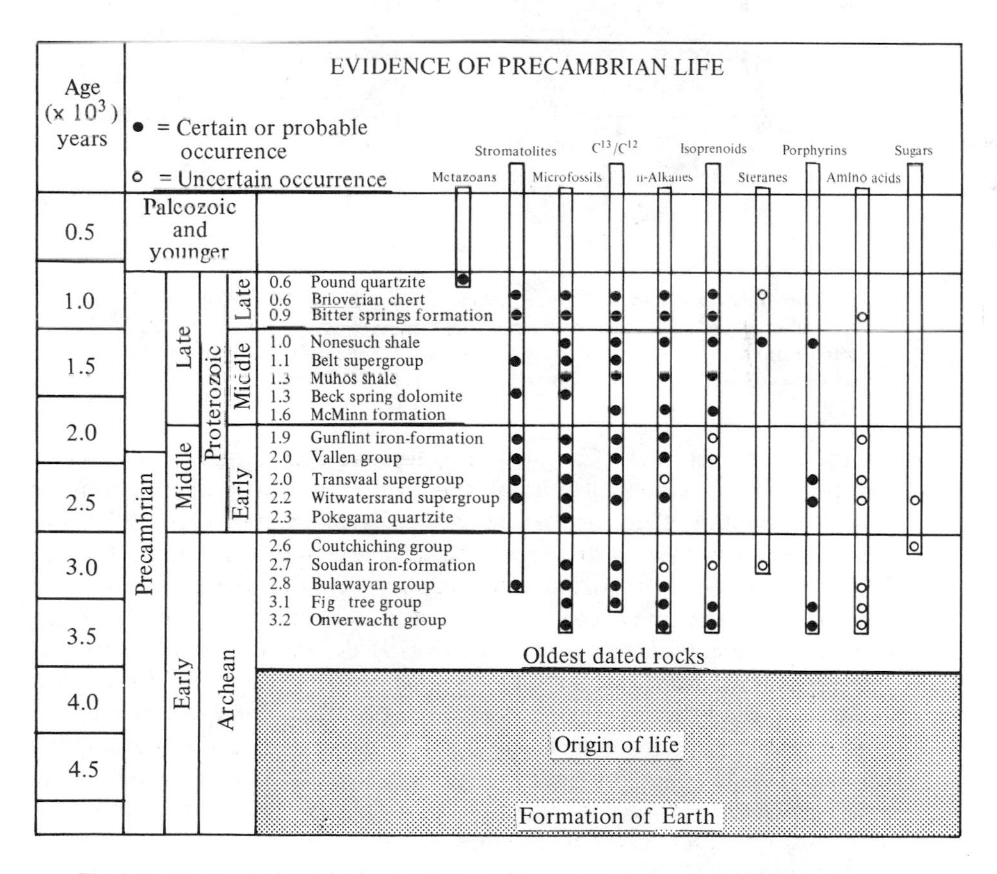

Fig. 3-5. Histogram showing the distribution of organic and morphological evidence of organisms in Precambrian sediments. (Adapted from Schopf, *Biological Reviews* **45**, 323, 1970.)

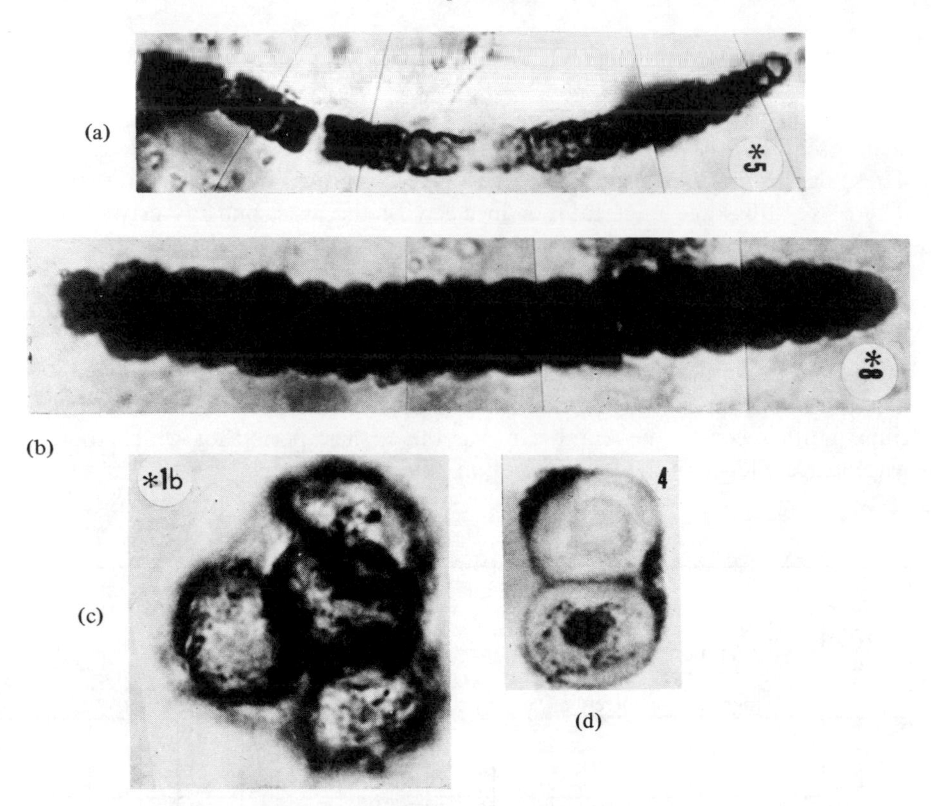

Fig. 3-6. Fossil algae from the Precambrian Bitter Springs Formation of Central Australia. (a) and (b) filamentous blue-green algae. (c) Eucaryotic algae (red algae?) in tetrahedral tetrad produced by meiotic division. (d) Spheroidal eucaryotic green algae which were killed just after cell division. (Courtesy J. W. Schopf.)

The period before the Cambrian is called the Precambrian. There are a number of schemes for subdividing this period, but no one scheme is in general use. Since there are no macrofossils in Precambrian rocks, we cannot easily determine a sequence of relative ages in the usual way. Hence, the rocks are usually classified according to their absolute ages or by the areas in which they are found. Precambrian rocks of a given type look the same despite great differences in age, and only the absolute dating methods can tell us which were formed first. In spite of the single name of this geological era, the Precambrian occupies about 85% of geological time. The Precambrian period has not received nearly as much attention from geologists as have later periods—partly because of the lack of fossils, and partly because oil, the commercial incentive for much geological work, does not occur often in Precambrian formations.

The absence of ordinary fossils does not necessarily mean the absence of evidence of life. Since about 1960 a limited number of fossils of microorganisms have been found in early Precambrian rocks and described in

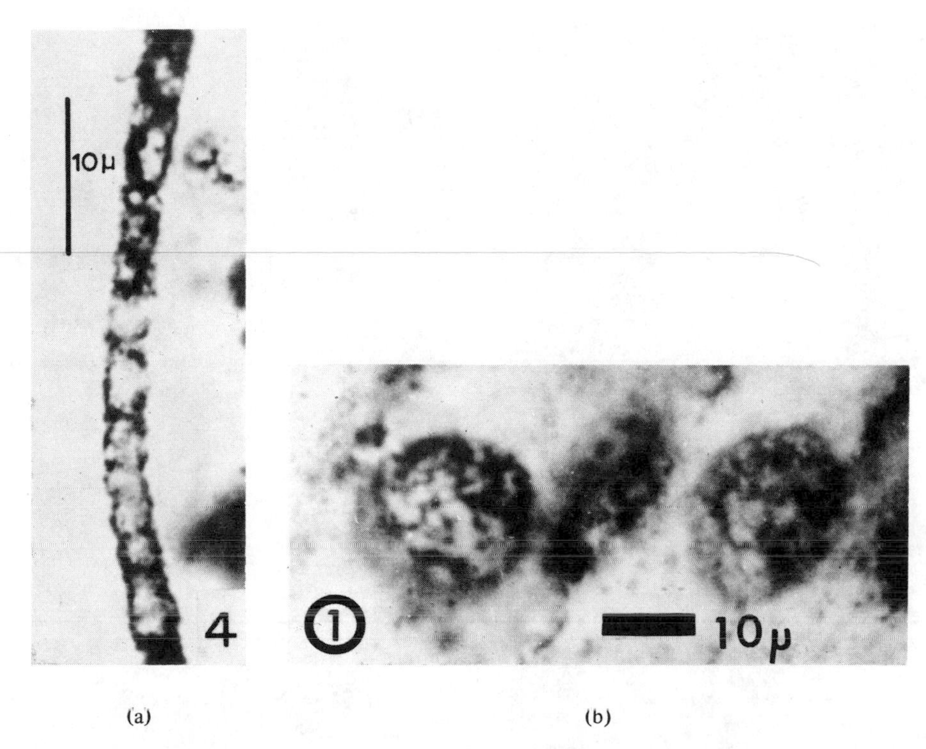

(a) (b)

Fig. 3-7. Septate algal filaments and reticulate algal-like spheroids in thin sections of the Middle Precambrian Gunflint chert. These are the most common microorganisms of the diverse Gunflint assemblage. (From Barghoorn and Tyler, *Science* **147**, 563, 1965.)

detail. Two factors made this work very difficult. Firstly, organisms without hard shells do not leave easily detectable fossils, although some evidence of them does remain. Secondly, very old rocks have usually been heated at one time or another, and heating destroys or greatly alters the organic remains. It may also distort the structural form of the fossil beyond recognition. The older a rock is, the more likely it is to have been heated at least once. Fortunately, there are a few Precambrian rocks which have not been heated, and a considerable amount of information has been obtained from them.

The principal Precambrian rocks that have been examined for their content of microfossils and organic compounds are

Bitter Springs chert (0.9×10^9 yr, Central Australia)
Nonesuch shale (1×10^9 yr, Northern Michigan)
Gunflint chert (1.9×10^9 yr, Southern Ontario)
Soudan shale (2.7×10^9 yr, Northeastern Minnesota)
Bulawayan limestone (2.7×10^9 yr, Rhodesia)
Fig-Tree chert (3.1×10^9 yr, Transvaal, South Africa)
Onverwacht (3.2×10^9 yr, Transvaal, South Africa)

Fig. 3-8. Precambrian stromatolites. (Top left photograph from William Schopf. Right photograph from McAlester, *The History of Life*, Prentice-Hall, Inc., Englewood Cliffs, New Jersey, 1971.)

The quaint names of these rocks have been given by geologists and refer to the formation in which the rock occurs. Thus, the Gunflint formation is so named because it contains a considerable amount of flint that was used for flintlock rifles. The Fig-Tree series is named after the town of Fig Tree, which is close to the formation.

There are no rocks known that are 4.5×10^9 or even 4.0×10^9 years old, so the geological record cannot yet tell us anything about conditions on the earth so long ago. Although some rocks of 3.5×10^9 years have been reported, the Fig-Tree and Onverwacht series are about as old as any rocks that are reliably dated. Geologists are currently looking for older rocks; if they are successful, they may be able to show that life began earlier than we now know. They might also learn about the atmospheric conditions on the

Fig. 3-9. Bulawayan limestone. Reduced to 70% of actual size. (From J. William Schopf.)

earth at the time when these earliest rocks were laid down. Of course, care would have to be exercised in such interpretations, since individual rocks may not have been formed under the conditions representative of the earth as a whole.

Well-preserved fossils of both algae and bacteria have been identified in the Gunflint chert; the Bulawayan limestone is said to be a fossil algal colony, looking very similar to the limestones being laid down off the Bahamas at the present time; the Fig-Tree chert contains fossils similar in size and form to bacteria. Several of these microfossils are shown in Fig. 3-10. No microfossils have yet been found in the Nonesuch shale or the Soudan shale. However, the absence of microfossils does not mean that life was absent at the time and place where these rocks were laid down, since it requires special circumstances to preserve microfossils.

It is interesting that the earliest known fossil microorganisms, for example those which occur in the Fig-Tree and Onverwacht cherts, resemble modern bacteria. They are called *prokaryotic* organisms because they have no nuclei. The organisms in the Gunflint chert resemble blue-green algae, which are also prokaryotic. The organisms in the more recent Bitter Springs chert are more complicated and resemble green algae, which have a well-defined nucleus and hence are termed *eukaryotic.*

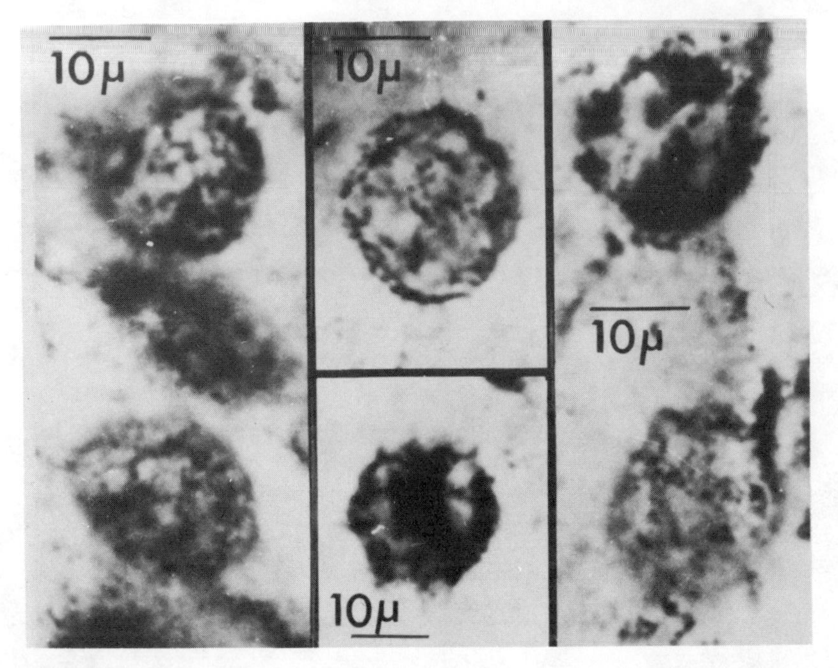

Fig. 3-10. Organic spheroidal bodies in thin sections of the Fig Tree chert. These microfossils appear to be the remnants of early Precambrian photosynthetic blue-green algae. (From Schopf and Barghoorn, *Science* **156**, 508, 1967.)

In addition to microfossils these rocks contain organic materials. Some of the compounds that have been identified are:

Nonesuch shale	hydrocarbons, pristane, phytane, porphyrins
Gunflint chert	normal paraffins, pristane, phytane
Soudan shale	alkanes, pristane, phytane, other isoprenoids
Bulawayan limestone	hydrocarbons, fatty acids
Fig-Tree chert	hydrocarbons, pristane, and phytane

The presence of organic compounds suggests that living organisms were present when the rock was laid down. The detection of porphyrins in the Nonesuch shale indicates that photosynthetic organisms were probably present on the earth one billion years ago. The pristane and phytane in the Soudan and Fig-Tree cherts provide further evidence of photosynthetic organisms as much as 3.1×10^9 years old, since the isoprenoid side chains of chlorophyll are the obvious source of these two compounds. However, since isoprenoid chains also occur in the fatty acids of halophilic organisms—organisms that live in concentrated salt solution—these organisms, rather than photosynthetic algae, could possibly be the source of the pristane and phytane.

(a)

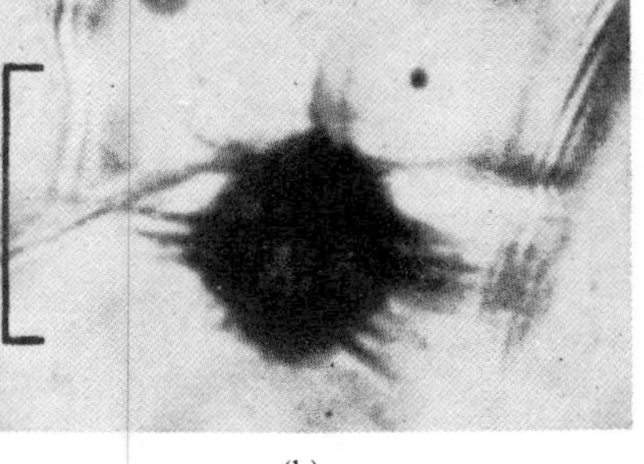

(b)

Fig. 3-11. (a) Pseudofossils—objects that resemble but are not true fossils. On the right is a pebble shaped by stream action to resemble an egg. On the left is a dendrite, a fern-like but entirely inorganic growth of manganese minerals that has spread along a narrow fracture in a rock. A "moss-agate" is the same type of structure, with the mineral matter intruding narrow cracks in the transparent material. (From Stokes, *Essentials of Earth History*, Prentice-Hall, Inc., Englewood Cliffs, New Jersey, 1966.) (b) Pseudofossil observed in a microbreccia sample from the Apollo 11 lunar dust samples. The pseudofossil was apparently formed by partial devitrification of the surrounding glassy matrix. The scale represents 10 μm. (From Schopf, *Science* **167**, 779, 1970.)

The interpretation of these results is not without uncertainties. One problem concerns the genuineness of the microfossils. While they are generally accepted as authentic, we cannot completely rule out the possibility of inorganic artifacts. Thus many colloidal materials, for example ferric hydroxide after it has precipitated and dried out, can form organized elements that look like organisms. Artifacts composed of iron, silica, and other materials could certainly occur in ancient sediments.

Again, the Bulawayan limestone is said to be a fossil algal colony which looks very similar to the algal colonies being laid down today off the Bahamas. However, Bahamian limestones are formed by *both* biological and nonbiological precipitation of $CaCO_3$. It is thus possible, although not likely, that the Bulawayan limestone could have been formed nonbiologically.

Next we must ask whether the organic compounds found in these rocks could be contaminants. Contamination can occur during handling, although this does not seem likely in the cases discussed above. However, contamination frequently occurs by absorption of organic materials from ground waters passing through rock formations. When there is ground water beneath a rock, it may be forced through the overlying rock if the pressure is great enough. Dissolved organic material can then be absorbed by the rock. This frequently occurs with limestones, which are quite porous. The flow of water through a chert is much slower, and it is generally felt that the Fig-Tree chert, for example, has not been flushed with water since it was laid down. This, however, has not been proved.*

The most important information derived from these geological data is the time scale for the origin of life. Since, despite the above reservations, the evidence seems fairly good that there was life on earth at the time the Fig-Tree was laid down 3.1×10^9 years ago, the origin of life must have occurred between 4.6 and 3.1×10^9 years ago. This is a period of 1.5×10^9 years. The life forms in the Fig-Tree chert, while not advanced types, must have required considerable time for their evolution from the most primitive organisms. A period of several hundred million years for this evolution is not unreasonable, so it seems almost certain that the first living organism arose not much later than 3.6×10^9 years ago, or about 1.0×10^9 years after the formation of the earth.

There is no way at present to estimate when, during this billion or so years, life arose. Many writers have stressed that many improbable events were required before life could begin and hence that much time was needed. They draw comfort from the billion years which seem to have been available. In general, we believe that such emphasis is misplaced. Periods of a hundred million years are so far removed from our experience that we can

* Recent evidence (J. W. Smith, J. W. Schopf, and I. R. Kaplan, *Geochim. Cosmochim. Acta* **34**, 659 (1970)) shows that in a number of Precambrian rocks, organic materials have been brought in recently by ground waters. This casts considerable doubt on claims to have found early Precambrian organic compounds in association with microfossils.

have no feeling or judgment as to what is likely or unlikely, probable or improbable, within them. If the formation of the first living organism took only one million years, we would not be very surprised. We cannot even prove that 10,000 years is too short a period.

There are further difficulties with all theories that postulate that the origin of life required a very long time. The organic compounds synthesized in the primitive atmosphere and oceans would have decomposed with time in many different ways. Although organic compounds could have been resynthesized from the decomposition products if the conditions were favorable, the loss of organic compounds would have been irrevocable if the atmosphere turned too oxidizing. This would certainly have happened as soon as molecular oxygen appeared in the atmosphere, but it could possibly have happened at an earlier stage when all the carbon in the atmosphere had been converted to CO_2. We cannot estimate when the atmosphere turned "too oxidizing" for organic synthesis. However, it is possible that conditions on the earth became less favorable for the synthesis of organic compounds even before the biological accumulation of oxygen.

These ideas have a more general application. There is a period in the evolution of earth-like planets when organic compounds are abundant. If life arises and becomes established during this period, it can adapt to later changes in the composition of the atmosphere. However, if life does not become established soon enough, the organic compounds on the surface of the planet decompose and can no longer be replaced. Once this has happened, the planet must remain lifeless forever.

In summary, the geological data show that life arose sometime within about a billion years of the formation of the earth. A more precise estimate of the time needed for the synthesis of the first living organisms cannot yet be given.

References

Geological dating

H. Faul, *Ages of Rocks, Planets and Stars* (McGraw-Hill, New York, 1966). An elementary discussion.

D. L. Eicher, *Geologic Time* (Prentice-Hall, Englewood Cliffs, N.J., 1968). An elementary discussion.

E. I. Hamilton, *Applied Geochronology* (Academic Press, London, 1965). A thorough treatment of all the radioactive dating methods.

Review articles on Precambrian microfossils

J. W. Schopf, "Pre-Cambrian micro-fossils and evolutionary events prior to the origin of vascular plants," *Biological Reviews* **45**, 319 (1970).

J. W. Schopf, "Antiquity and evolution of Pre Cambrian life," *McGraw-Hill Yearbook of Science and Technology*, 1967, p. 47.

Precambrian fossils: original papers

E. S. Barghoorn, "Origin of life," in *Treatise on Marine Ecology and Paleoecology, Geol. Soc. Amer. Mem.* 67 (Vol. 2), 75–85 (1957).

E. S. Barghoorn and S. A. Tyler, *Ann. N.Y. Acad. Sci.* **108**, 451 (1963).

E. S. Barghoorn and S. A. Tyler, *Science* **147**, 563 (1965) (Gunflint Chert).

E. S. Barghoorn, W. G. Meinschein, and J. W. Schopf, *Science* **148**, 461 (1965).

E. S. Barghoorn and J. W. Schopf, *Science* **150**, 337 (1965) (Bitter Springs).

E. S. Barghoorn and J. W. Schopf, *Science* **152**, 758 (1966) (Fig-Tree Chert).

J. W. Schopf, E. S. Barghoorn, M. D. Maser, and R. Gordon, *Science* **149**, 1365 (1965) (Fig Tree).

J. W. Schopf and E. S. Barghoorn, *Science* **156**, 508 (1967) (Fig Tree).

J. W. Schopf, K. A. Kvenvolden, and E. S. Barghoorn, *Proc. Nat. Acad. Sci. U.S.* **59**, 639 (1968) (Amino Acids in Precambrian Sediments).

A. E. Engel, B. Nagy, L. A. Nagy, C. G. Engel, C. W. W. Kremp, and C. M. Drew, *Science* **161**, 1005 (1968) (Onverwacht).

Reviews on organic geochemistry of Precambrian rocks

M. Calvin, *Chemical Evolution—Molecular Evolution Towards the Origin of Living Systems on the Earth and Elsewhere* (Oxford University Press, New York, 1969).

G. Eglinton, "Hydrocarbons and fatty acids in living organisms and recent and ancient sediments," in P. A. Schenck and I. Havenaar, eds., *Advances in Organic Geochemistry* (Pergamon Press, London, 1968), pp. 1–24.

G. Eglinton, P. M. Scott, T. Belsky, A. L. Burlingame, W. Richter, and M. Calvin, "Occurrence of Isoprenoid Alkanes in a Precambrian Sediment," in *Advances in Organic Geochemistry*, G. D. Hobson and M. C. Louis, eds. (Pergamon Press, Oxford, 1966), pp. 41–74.

G. Eglinton and M. Calvin, "Chemical fossils," *Sci. Amer.* **216**, 32 (Jan. 1967).

P. H. Abelson, "Geochemistry of Organic Substances," in *Researches in Geochemistry*, P. H. Abelson, ed. (John Wiley, New York, 1959), pp. 79–103.

I. A. Breger, ed., *Organic Geochemistry* (Macmillan, New York, 1963).

Chapter Four

The Composition of the Primitive Atmosphere

INTRODUCTION

Geological and geophysical evidence is insufficient to allow us to state with any precision what conditions were like on the surface of the primitive earth. Arguments concerning the composition of the primitive atmosphere are particularly controversial. It is important, therefore, to state our own prejudice clearly. We believe that there must have been a period when the earth's atmosphere was reducing, because the synthesis of compounds of biological interest takes place only under reducing conditions.

There is, indeed, some geological and geophysical evidence which points in the same direction, but none of it is conclusive. Many authors have argued that since the abundance of hydrogen in the universe is so much greater than that of any other element, the earth's atmosphere must have been reducing. However, the atmosphere is now thought to have arisen by degassing of the interior. It is at least possible that the earliest volcanic gases contained very little molecular hydrogen and that carbon dioxide rather than carbon monoxide or methane was in excess. Ferrous iron—unstable in the presence of O_2—is found in early Precambrian formations; therefore, the early Precambrian atmosphere is said to have been free of O_2. However, this does not prove that it was reducing. Fortunately, everyone agrees that although the primitive atmosphere may not have been strongly reducing, it certainly did not contain more than a trace of molecular oxygen.

We shall first discuss the contemporary atmosphere, which is strongly oxidizing. Then we shall consider the equilibrium forms of carbon and nitrogen in a reducing atmosphere at a moderate temperature and show that methane, nitrogen, and ammonia predominate. This leads naturally to a discussion of the loss of hydrogen which must have accompanied the transition from the primitive reducing atmosphere to the contemporary oxidizing atmosphere. Finally, we shall consider the complicated interaction of the atmosphere with the ocean, and give a speculative "history" of the oceans and atmosphere.

THE PRESENT OXIDIZING ATMOSPHERE

The composition of the present atmosphere is given in Table 4-1. The most abundant species is nitrogen and the next most abundant is oxygen. The principle characteristics of the atmosphere are determined by the presence of free oxygen.

Our atmosphere is not thermodynamically stable. The equilibrium constant for the reaction

$$\tfrac{1}{2}N_2 + \tfrac{1}{2}O_2 + \tfrac{1}{2}H_2O(liq) \rightleftarrows HNO_3(aq) \qquad K = \frac{(HNO_3)}{(N_2)^{1/2}(O_2)^{5/4}}$$

is $K = 0.04$ at 25°C. If the atmosphere were to equilibrate, considerable O_2 would react with N_2 to give HNO_3. The HNO_3 would dissolve in the ocean to give approximately a 0.01 M solution of nitric acid.

The formation of nitric acid is very slow at room temperature, but it would probably occur in geological times. In fact, considerable quantities of nitrate are produced in thunderstorms. The formation of nitric acid does not go to equilibrium because nitrate is reduced to molecular nitrogen and ammonia, mainly by living organisms.

Table 4-1. Composition of Clean Dry Air

	Volume or Mole %	g/cm²
N_2	78.09	757.4
O_2	20.95	232.2
A	0.93	12.8
CO_2	0.03	0.46
Ne	18×10^{-4}	12.5×10^{-3}
He	5.2×10^{-4}	0.72×10^{-3}
CH_4	1.5×10^{-4}	0.94×10^{-3}
Kr	1×10^{-4}	2.9×10^{-3}
N_2O	0.5×10^{-4}	0.8×10^{-3}
H_2	0.5×10^{-4}	0.035×10^{-3}
Xe	8×10^{-6}	0.36×10^{-3}

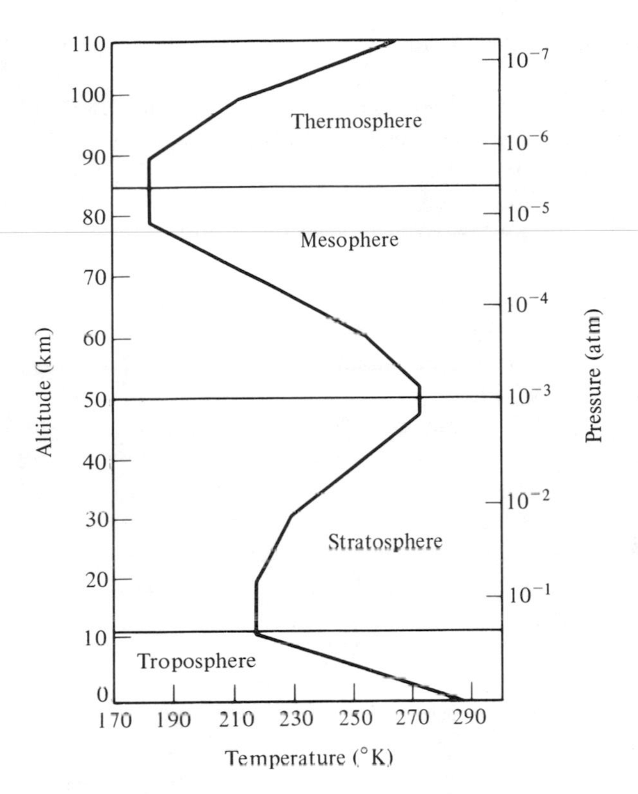

Fig. 4-1. Temperature variations in the atmosphere at altitudes below 110 km at a latitude of approximately 45°. The temperature falls with increasing height from sea level to the tropopause because the atmosphere is approximately adiobatic. At greater heights the variation in temperature is complicated because of the absorption and emission of infrared radiation. Above 110 km, the temperature rises rapidly with altitude—reaching approximately 1,500° at the escape layer at 600 km.

Hydrogen and methane are present in amounts far above their equilibrium values. Equilibrium constants for the relevant oxidation reactions are:

$$H_2 + \tfrac{1}{2}O_2 \leftrightarrows H_2O(\text{liq}) \qquad K_{25} = 4 \times 10^{41}$$

$$CH_4 + 2O_2 \leftrightarrows CO_2 + 2H_2O(\text{liq}) \qquad K_{25} = 2 \times 10^{143}$$

The equilibrium pressures are $p_{H_2} = 6 \times 10^{-42}$ atm and $p_{CH_4} = 10^{-146}$ atm while the observed values are 5×10^{-7} and 1.5×10^{-6} atm, respectively.

How do we account for such large deviations from equilibrium? Hydrogen is continuously produced in the upper atmosphere by photodissociation of water. Methane is produced in the ungulates (e.g., cows) and in the soil by methane bacteria. Both these gases at the concentrations present in the atmosphere react slowly with oxygen so that their steady-state concentrations can remain above the equilibrium values. Mixtures with air of H_2 or CH_4 at higher concentrations are explosive.

Here we have touched on a very difficult problem. It is simple to calculate the equilibrium concentrations in a chemical system (provided the thermodynamic data are available), but the only way to predict how closely equilibrium will be approached is by making kinetic measurements. This is difficult even for the hydrogen-oxygen reaction in the atmosphere, because it is not easy to allow for the effect of ultraviolet light.

It is often safe to assume that the principle components of a planetary atmosphere, those representing 1% of the atmosphere or more, are in approximate equilibrium. Consider the addition of 0.001 atm of NH_3 to an atmosphere containing 0.8 atm of N_2, 0.1 atm of NH_3, and enough H_2 to ensure equilibrium (3×10^{-3} atm at 25°C): The change in NH_3 concentration would be 1%. Next consider the addition of the same quantity of NH_3 to an equilibrium atmosphere containing 0.8 atm of N_2, 1×10^{-4} atm of NH_3 and 3×10^{-5} atm of H_2: This addition would increase the NH_3 concentration by a factor of eleven. Clearly, a slow addition of NH_3, say 0.001 atm per year, would be proportionally more important if the initial NH_3 concentration were low.

This argument purports to prove that planetary atmospheres are in approximate equilibrium. Our own atmosphere is an important exception to this rule, for we have seen that at equilibrium free oxygen would disappear and would be replaced by nitrate. Thus, the presence of living organisms can affect profoundly the composition of an atmosphere. This would be a good "symptom" of life if we could be sure that there are no other processes that can drive an atmosphere very far from equilibrium. Unfortunately, there may be photochemical processes which have the same effect.

THE REDUCING ATMOSPHERE IN THE ABSENCE OF AN OCEAN

The stable form of carbon in the presence of excess hydrogen at room temperature is methane. This is shown by the following equilibria:

$$CO_2 + 4H_2 \leftrightarrows CH_4 + 2H_2O(liq) \qquad K_{25} = 8.1 \times 10^{22} \qquad (1)$$

$$CO + 3H_2 \leftrightarrows CH_4 + H_2O(liq) \qquad K_{25} = 2.5 \times 10^{26} \qquad (2)$$

$$C + 2H_2 \leftrightarrows CH_4 \qquad K_{25} = 7.9 \times 10^{8} \qquad (3)$$

All intermediate oxidation states of carbon, except graphite and nonvolatile graphite-like materials, are unstable with respect to disproportionation to CH_4 and CO_2.

$$4HCO_2H(g) \leftrightarrows CH_4 + 3CO_2 + 2H_2O(g) \qquad K_{25^\circ} = 1.0 \times 10^{61}$$

$$2H_2CO(g) \leftrightarrows CH_4 + CO_2 \qquad K_{25°} = 4.0 \times 10^{39}$$

$$4CH_3OH(g) \leftrightarrows 3CH_4 + CO_2 + 2H_2O(g) \qquad K_{25°} = 2.6 \times 10^{62}$$

Thus, in any equilibrated atmosphere at moderate temperatures the only carbon compounds are CO_2 and CH_4; these compounds coexist in equilibrium. Of course, very small amounts of substances in intermediate oxidation states will be present. For example, if $p_{CO_2} = 1$ atm, $p_{CH_4} = 1$ atm, and $p_{H_2O} = 1$ atm, then $p_{H_2CO} = 1.6 \times 10^{-20}$ atm and $p_{CH_3OH} = 2.5 \times 10^{-16}$ atm.

All these equilibrium constants are temperature dependent. Reactions 1, 2, and 3 become less favorable at higher temperatures, and the equilibria are reversed at about 600°C so that carbon monoxide, for example, becomes stable relative to methane and water. However, the stability of *organic* compounds is not significantly increased at higher temperatures. More elaborate equilibrium calculations have been carried out, but these calculations show that all volatile organic compounds other than methane are unstable in both reducing and oxidizing atmospheres, especially in the latter.

In a reducing atmosphere, ammonia is the stable nitrogen-containing species. The synthesis of NH_3 from N_2 and H_2 is a well-known reaction and is used to produce millions of tons of NH_3 for fertilizer. The equilibrium constant is given by

$$\tfrac{1}{2}N_2 + \tfrac{3}{2}H_2 \leftrightarrows NH_3 \qquad K_{25} = 8.2 \times 10^2 \tag{4}$$

Note that at a given pressure of hydrogen the reaction $N_2 + 3H_2 = 2NH_3$ is not as favorable as the reaction of carbon with hydrogen to give methane. At higher temperatures the stability of NH_3 relative to N_2 and H_2 decreases.

THE ESCAPE OF HYDROGEN FROM THE ATMOSPHERE

For reasons already given, we believe that the earth's atmosphere was once reducing and that the present oxidizing atmosphere was formed as a consequence of the escape of hydrogen. Calculations show that the temperature and gravitational field of the earth are such that hydrogen and helium can escape at a significant rate, but that oxygen, nitrogen and heavier atoms cannot.

Early calculations by Jeans give the rate of escape L (in atoms cm^{-2} sec^{-1}) as

$$L = N\left(\frac{RT}{2\pi\mu}\right)^{1/2}(1 + x)e^{-x}, \qquad x = \frac{GM\mu}{RTa_c}$$

where N = atoms cm^{-3} in the escape layer, R = the gas constant, T is the absolute temperature in the escape layer, μ = the atomic weight of the gas,

G — constant of gravitation, M — Mass of earth, a_c = radius at the escape layer. This calculation assumes that the rate limiting step is the acquisition by an atom in the high atmosphere of enough radial velocity to escape the earth's gravitational field. It is applicable only in regions where the atmosphere is thin enough for an atom to escape without collision. The earth's escape layer, for example, begins about 600 km above the surface.

The exponential form of this equation makes the rate of escape from a planet very sensitive to the weight of the molecule, the mass of the planet, and the temperature and radius of the escape layer. Since the factors in the exponential are very different for other planets, the rates of escape will vary widely from planet to planet.

It is now doubtful whether Jeans' formula, or any of the various modifications of it, allows the calculation of the rate of escape from the earth's atmosphere. It appears that the rate limiting step in the escape of hydrogen from the earth is the diffusion of hydrogen-containing compounds from the lower atmosphere to the escape layer. The temperature of the escape layer is so high (1,500°K) that once hydrogen gets there, it almost certainly escapes. Any water or methane that gets to the escape layer undergoes photodissociation, and the hydrogen atoms so formed escape.

Hydrogen can diffuse to the escape layer in the form of H_2 (0.5×10^{-4} mole %), CH_4 (1.5×10^{-4} mole %), and H_2O. The water content of air varies from very low values to as much as 7 mole %, but the passage of water to the upper atmosphere is limited by a cold trap at the tropopause. The tropopause varies in height and temperature, being at 18 km and 186°K near the equator and at 7 km and 220°K near the poles. Since the air at the equator rises through the tropopause, travels to the poles in the stratosphere, and sinks at the poles through the tropopause, water picked up close to the earth is frozen out at the equatorial tropopause at 186°K. The vapor pressure of water at this temperature is 1.6×10^{-7} atm; the total atmospheric pressure at the equatorial tropopause is 7.9×10^{-2} atm. Thus, despite its abundance in the lower atmosphere, water vapor makes up only 2.0×10^{-4} mole % of the air at the tropopause, a proportion comparable to those of H_2 and CH_4. It appears that water is not the only important source of hydrogen in the escape layer.

When allowance is made for these complications, it is estimated that 10^7 atoms cm^{-2} sec^{-1} of hydrogen are escaping from our atmosphere. If this rate has applied since the formation of the earth, only the hydrogen equivalent to 20 g cm^{-2} of water would have escaped in 4.5×10^9 years. This is insufficient to account for the oxygen in the atmosphere (230 g cm^{-2}). Thus it seems likely that, in the past, much more hydrogen has reached the escape layer as H_2.

We can make use of the rate of escape of hydrogen to estimate the partial pressures of hydrogen (p_{H_2}) on the primitive earth, if we assume that carbon, nitrogen, oxygen, and sulfur were in a reduced form when the earth was formed. Since they are now largely oxidized, we suppose that the hydrogen released during their oxidation must have escaped from the atmosphere. Hydrogen must also have been released during the oxidation of FeO to Fe_2O_3. Table 4-2 summarizes the quantities of hydrogen involved. The largest source of hydrogen must have been reduced carbon. The value given in Table 4-2 assumes that carbon on the primitive earth was present initially as graphite or as carbides. If it was retained as CH_4, the figure of 9,000 g cm^{-2} would be doubled.

Table 4-2. Loss of Hydrogen from the Oxidation of Reduced Species

Element Oxidized	g cm^{-2} of H_2O Equivalent to this Hydrogen
Oxygen in atmosphere (230 g cm^{-2}) $H_2O \rightarrow H_2 + \frac{1}{2}O_2$	260
Surface Carbon (3,000 g cm^{-2}) $C + 2H_2O \rightarrow 2H_2 + CO_2$	9,000
Nitrogen in Atmosphere (755 g cm^{-2}) $NH_3 \rightarrow \frac{3}{2}H_2 + \frac{1}{2}N_2$	1,460
Sulfate in sea water (245 g cm^{-2}) $S + 4H_2O \rightarrow 3H_2 + H_2SO_4$	410
Ferric iron in weathered rocks (1,000 g cm^{-2}) $2FeO + H_2O \rightarrow H_2 + Fe_2O_3$	450
TOTAL	11,580

The hydrogen pressure must have been quite high to permit the escape of hydrogen equivalent to 12,500 g cm^{-2} of water. We assume that the escape of H_2 was limited by its diffusion to the escape layer. If the hydrogen in 20 g of water can escape in 4.5×10^9 years when the pressure of hydrogen is 0.5×10^{-6} atm, then 625 times that pressure of hydrogen would have been necessary for the escape of the hydrogen in 12,500 g cm^{-2} of water (that is, 3.1×10^{-4} atm H_2). If the escape occurred over a period of 1×10^9 years, the pressure of hydrogen must have been 1.4×10^{-3} atm.

This estimate of H_2 pressure depends on two further assumptions beyond the assumption of an initially fully reduced atmosphere. First, we have supposed that the limiting step has always been transport to the escape layer rather than the final escape. Second, we have assumed that hydrogen reached the escape layer as molecular hydrogen rather than ammonia or methane. We are not certain that either assumption is correct for the primitive earth. However, we believe our estimates should be correct to about an order of magnitude; that is, we believe that the hydrogen pressure was between 10^{-4} and 10^{-2} atm.

THE PRIMITIVE ATMOSPHERE AND OCEAN AND THEIR INTERACTION

Older textbooks state that the pH of the ocean is controlled by the bicarbonate concentration and the carbon dioxide pressure. This view is no longer held. The pH of the present ocean varies between 8.0 and 8.5; we will use the value of 8.1 in our discussion. The pH, Ca^{++}, Mg^{++}, Na^+, and K^+ concentrations are all believed to be fixed by rather poorly understood equilibria involving clay minerals. The bicarbonate concentration and carbon dioxide pressure are determined by the pH and the Ca^+ concentration.

Examples of the many important silicate and clay mineral equilibria are:

$$\underset{\textbf{Muscovite}}{2H^+ + 2KAl_3Si_3O_{10}(OH)_2} + 3H_2O \leftrightharpoons 2K^+ + \underset{\textbf{Kaolinite}}{3Al_2Si_2O_5(OH)_4} \qquad (5)$$

$$2H^+ + \underset{\textbf{Montmorillonite}}{6Na_{0.33}Al_{2.33}Si_{3.67}O_{10}(OH)_2} + 7H_2O \leftrightharpoons 2Na^+ + \underset{\textbf{Kaolinite}}{7Al_2Si_2O_5(OH)_4} + 8SiO_2 \qquad (6)$$

$$10H^+ + \underset{\textbf{Chlorite}}{Mg_5Al_2Si_3O_{10}(OH)_8} \leftrightharpoons 5Mg^{++} + \underset{\textbf{Kaolinite}}{Al_2Si_2O_5(OH)_4} + SiO_2 + 7H_2O \qquad (7)$$

The above formulae are those of pure minerals, but in nature each pure mineral absorbs considerable quantities of foreign ions. Of these, Ca^{++} is particularly important. If the pH of ocean were to fall sufficiently, the equilibria would be pushed to the right and the clay minerals on the left would be converted to those on the right. These equilibria would provide a crude buffer mechanism for the oceans.

Superimposed on this coarse buffer action, a pH control is caused by ion exchange on clay minerals, in this case without change in the mineral phase.

$$2H^+ + Clay \cdot Ca^{++} \leftrightharpoons Ca^{++} + Clay \cdot 2H^+ \qquad (8)$$

We can also write ion exchange equilibria for the other ions.

$$H^+ + Clay \cdot Na^+ \leftrightharpoons Na^+ + Clay \cdot H^+ \qquad (9)$$

Table 4-3. The Concentrations of the Major Ions in Sea Water

	Molarity		Molarity
Na^+	0.481	Cl^-	0.561
Mg^{++}	0.055	$SO_4^=$	0.029
Ca^{++}	0.0104	HCO_3^-	0.0024
K^+	0.0102	$CO_3^=$	0.0003
Sr^{++}	0.0001	Br^-	0.0008
		$HPO_4^= + H_2PO_4^-$	$\leq 3 \times 10^{-6}$

$$H^+ + \text{Clay} \cdot K^+ \leftrightarrows K^+ + \text{Clay} \cdot H^+ \tag{10}$$

If the H^+ concentration is fixed by Equations 5, 6, and 7, or the exchange with K^+ and Na^+ in clay minerals (Equations 9 and 10), then the Ca^{++} concentration is fixed by Equation 8.

The actual situation is more complex than that described here, but the equations given above are believed to describe the basic process occurring in the present-day oceans. It seems reasonable that there was substantial weathering of continental rocks on the primitive earth, resulting in the formation of extensive oceanic sediments. If so, a similar set of equilibria must have controlled the pH and concentrations of Ca^{++} and other ions in the primitive ocean, thereby leading to similar values for the pH and Ca^{++}.

The partial pressure of CO_2—p_{CO_2}—in the present atmosphere is fixed by the equilibrium

$$CaCO_3 + 2H^+ \leftrightarrows Ca^{++} + CO_2 + H_2O$$
$$p_{CO_2} = \frac{2 \times 10^{10}(H^+)^2}{Ca^{++}} \tag{11}$$

Once the H^+ and Ca^{++} are known, p_{CO_2} can be calculated provided CO_2 is in equilibrium with solid $CaCO_3$. Substitution of the present value of the pH (8.1) and the Ca^{++} in the oceans (0.01 M) in Equation 11 gives $p_{CO_2} = 1.3 \times 10^{-4}$ atm. If allowance is made for the activity coefficient of Ca^{++}, the p_{CO_2} is calculated to be 3×10^{-4} atm. The HCO_3^- and $CO_3^=$ concentrations can then be calculated from the first and second dissociation constants of carbonic acid and the pH.

The above discussion examines only one part of the process of $CaCO_3$ precipitation and the control of the p_{CO_2} in the atmosphere. The calcium in the oceans comes from the weathering of silicate-rich rocks; the CO_2 comes from biological oxidation of organic compounds, weathering of carbonate rocks, and volcanoes. A simple equation that expresses this overall process is

$$\underset{\textbf{Wollastonite}}{CaSiO_3} + CO_2 \leftrightarrows CaCO_3 + \underset{\textbf{Quartz}}{SiO_2}$$

The equilibrium partial pressure of CO_2 for this reaction at 25°C is about 10^{-7} atm. This particularly simple reaction is not important, since Wollastonite is a rare mineral; but a number of related and more complex equilibria are believed to be important in controlling the p_{CO_2}. Examples are

$$\underset{\textbf{Anorthite}}{CaAl_2Si_2O_8} + CO_2 + 2H_2O \leftrightarrows CaCO_3 + \underset{\textbf{Kaolinite}}{Al_2Si_2O_5(OH)_4}$$

$$\underset{\textbf{Diopside}}{CaMg(SiO_3)_2} + CO_2 \leftrightarrows \underset{\textbf{Dolomite}}{CaMg(CO_3)_2} + \underset{\textbf{Quartz}}{SiO_2}$$

$$\underset{\textbf{Diopside}}{CaMg(SiO_3)_2} + CO_2 \leftrightarrows CaCO_3 + \text{Magnesium silicates}$$

Each of these reactions has a different equilibrium partial pressure of CO_2, and in addition, equilibrium is not reached in all these reactions. This results in a complicated and not well understood process. Nevertheless, it seems clear that the p_{CO_2} on the earth is kept low by these and related reactions, even though the exact p_{CO_2} cannot be calculated. This process is referred to as the *Urey Equilibrium.*

Nowadays, calcium carbonate is precipitated in large quantities over the earth, almost always by living organisms. In the absence of living organisms, precipitation might not occur until solutions became supersaturated with $CaCO_3$, perhaps until they contained twice the saturating concentration. This would only double the p_{CO_2} required to produce a precipitate. We see that, if $CaCO_3$ was precipitated on the primitive earth, the p_{CO_2} must have been about the same as at present.

There are not many early Precambrian limestones, and no rocks older than 3.5×10^9 years are known, so we cannot be sure about the presence of solid $CaCO_3$ on the primitive earth. However, since it does not require much limestone to establish equilibrium with Ca^{++} and CO_2, we think it likely (except possibly for a short period immediately after the earth became a solid body) that sufficient $CaCO_3$ was present.

A pH value of 8.1 estimated on geological grounds fits in very well with our ideas of prebiotic syntheses. Although some reactions go better at a little higher pH or a little lower, a pH of about 8 seems the best compromise for known prebiotic reactions.

EVOLUTION OF THE ATMOSPHERE

We are now ready to discuss the equilibrium of carbon species in the atmosphere-oceans-sediments system. We assume first that all the surface carbon was initially in the atmosphere, giving about 10 atm of CH_4. The hydrogen pressure is arbitrarily taken to have been 0.1 atm, and we suppose it to have decreased as hydrogen escaped from the atmosphere (see Fig. 4-2). With this model, p_{CO_2} was initially 10^{-18} atm and $CaCO_3$ could therefore not have precipitated. Methane was the principal carbon species on the primitive earth until the p_{H_2} dropped to 10^{-4} atm, whereupon graphite began to form. Methane and graphite together were stable until the p_{H_2} fell to 6×10^{-6} atm. At that p_{H_2}, the p_{CO_2} reached 3×10^{-4} atm and $CaCO_3$ began to deposit. At this time, the pressure of CH_4 was 0.03 atm, so it was still the most abundant carbon species in the atmosphere. When p_{H_2} fell further, graphite was no longer stable, and the methane was in equilibrium only

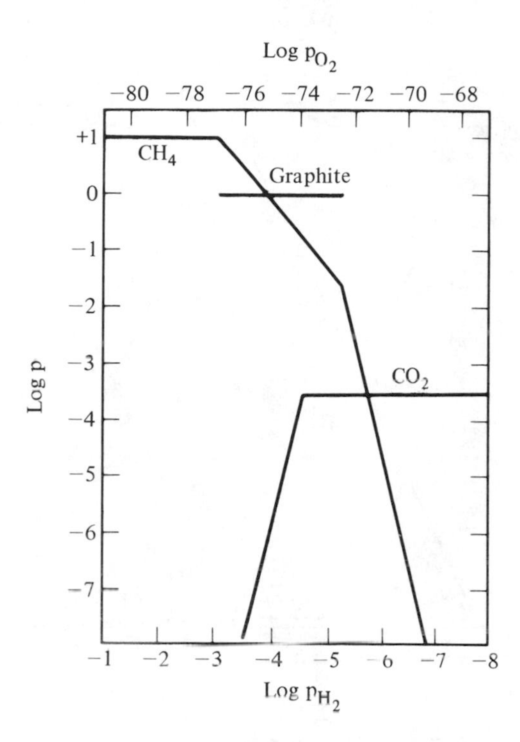

Fig. 4-2. The partial pressures of CH_4 and CO_2 as a function of p_{H_2} at equilibrium with both atmosphere and ocean. The graphite line corresponds to solid graphite (activity = 1).

with the $CaCO_3$. Methane remained the most abundant carbon species in the atmosphere until the p_{H_2} fell to 2×10^{-6} atm.

We next consider the time course of the conversion described above. In doing so, we again assume that all the carbon was initially in the atmosphere as CH_4 (10 atm + 0.1 atm H_2), and that this CH_4 was dehydrogenated to CO_2 in 1×10^9 years. We now make the additional assumption that the rate of escape of H_2 was constant.

The time course of methane decomposition is shown in Fig. 4-2. The chemical reaction for the first step is

$$CH_4 \rightarrow C + 2H_2$$

When all but 0.3% of the CH_4 had been converted to graphite, the graphite reacted with water

$$C + 2H_2O \rightarrow CO_2 + 2H_2$$

and the CO_2 was taken up as $CaCO_3$. The equilibrium p_{H_2} was constant during this reaction. During the decomposition of almost all the CH_4, the p_{H_2} was maintained in the relatively narrow range between 1×10^{-4} and 6×10^{-6} atm.

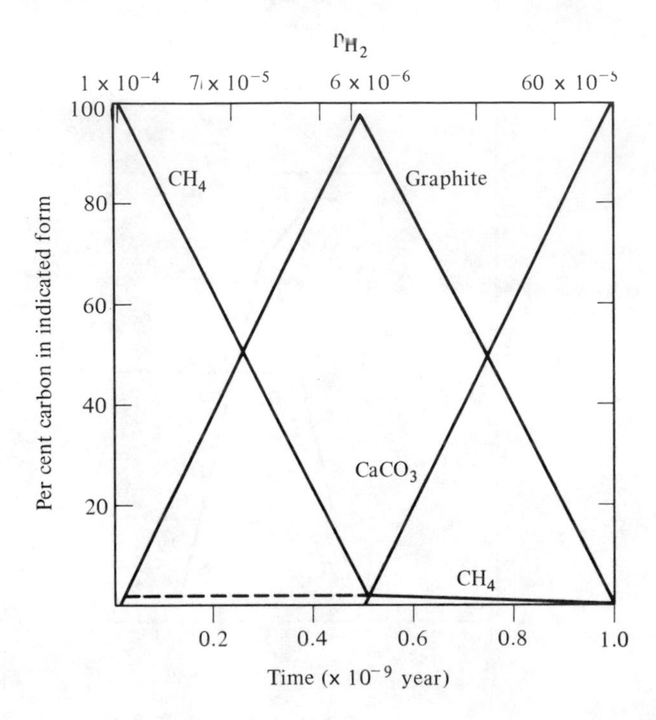

Fig. 4-3. The percent carbon in various forms as a function of time. Equilibrium is assumed and graphite is allowed to form. The escape of hydrogen is constant and is completed in 1×10^9 years. The solid CH_4 line is the percent carbon in the atmosphere when all the CH_4 is initially in the atmosphere. The dashed CH_4 line is the CH_4 in the atmosphere when carbon is added to the atmosphere as CH_4 over a period of 0.5 billion years by degassing the earth.

The assumption that all the surface carbon on the earth was present in the initial atmosphere is too extreme, since we think that much of it arose by degassing from the interior of the earth. It may, for example, have taken as much as 500 million years for the carbon to reach the surface. If we take account of this complication but assume that carbon degassed as methane, we must change our detailed schedule of carbon oxidation, but not our general conclusions.

If we make the more realistic assumption that much CO_2 and CO was present in the primitive volcanic gases, we must recognize that the earliest atmosphere may already have corresponded to an "evolved" atmosphere (as in Fig. 4-3, for example, by the situation marked as 0.2 or even 0.6×10^{-7} year). Fortunately, even great uncertainties here do not affect our general conclusions, since the organic syntheses on which our theories depend occur in any reduced atmosphere, even in an atmosphere containing a great deal of CO.

Since we assumed equilibria that could not have been attained, it seems very clear for several reasons that the discussion given above cannot be a complete account of the early history of the atmosphere. In particular:

1. The system (CH_4, CO_2, graphite, H_2O, and H_2) does not reach equilibrium rapidly at low temperature. In particular, it is not likely that graphite was formed in the amounts predicted by equilibrium calculations. Graphite does not nucleate readily from the gas phase, and it has appeared in only a few experiments in which mixtures containing methane are subjected to electron discharges or ultraviolet irradiation under primitive earth conditions.
2. Ultraviolet light and electric discharge, which must have been important on the primitive earth, act on a mixture of CH_4 and H_2O to form CO_2 and CO.
3. The proportions of CH_4, CO, and CO_2 at equilibrium are pressure dependent. Lowering the pressure increases the ratio of CO_2 to CH_4. If equilibrium was approximated at high altitude and the gas mixture brought to the surface of the earth, CO_2 and CO would be out of equilibrium and more abundant than predicted by equilibrium calculations for 1 atm total pressure.

It could be argued that kinetic factors are so important in the reactions of CH_4 and NH_3 that equilibrium considerations are not even a rough guide to the species present in the primitive atmosphere. This objection would be valid if there were no mechanism for synthesis of CH_4 from, say, CO_2, CO, and H_2. The synthesis of methane from H_2 and CO_2 or CO is slow in the absence of catalysts at 25° or 100°C, but it is likely that there are radical reactions that produce CH_4 from more oxidized species at low temperatures. More work is needed to resolve this problem.

The atmospheres of the major planets contain CH_4. Presumably it could have been present in the earth's atmosphere in the presence of sufficient H_2, provided the methane was not destroyed by the greater intensity of ultraviolet light.

These kinetic considerations lead us to the conclusion that CO_2 was more abundant than predicted by equilibrium calculations. However, the reaction of CO_2 with Ca^{++} is rapid enough to make it unlikely that the CO_2 in the atmosphere could have risen much above the present partial pressure of 3×10^{-4} atm. Small amounts of $CaCO_3$ were probably precipitated sooner than is suggested by Fig. 4-2. The carbon monoxide concentration was also probably higher than is suggested by equilibrium calculations (see the following section for a discussion of CO).

The most important deviation from equilibrium must have involved organic compounds. As we shall show in Chapter 7, organic compounds are readily synthesized in a "primitive-earth" atmosphere. Once in the oceans, they are kinetically stable relative to decomposition to graphite, and hence, we believe, they accumulated in large amounts. Thus the line "graphite" in Fig. 4-2 might well be replaced by "mainly organic compounds." We believe that large quantities of organic material were formed on the primitive earth, even though they would not have been present in equilibrium.

An example of how nonequilibrium conditions might have occurred in a reducing atmosphere is the proposal that the ocean was covered by an oil slick several meters thick. The hydrocarbons of this oil slick are supposed to have been formed by photochemical decomposition of methane by light of wavelengths below 1,450 Å at the 100 to 300 km level of the atmosphere. The methyl radicals from the photodissociation of methane would have built up into higher hydrocarbons until their vapor pressures were so low that they would precipitate out of the atmosphere. Since the solubility and density of hydrocarbons are low, they would have accumulated on the surface of the ocean. Experiments have been performed that show that higher hydrocarbons can be formed from methane by short-wavelength ultraviolet light.

We are not convinced that an oil slick was as important as this proposal indicates, since polar groups (e.g., H_2O and NH_3) would be added to the hydrocarbons during the photochemical reactions in the atmosphere. Hydrocarbons with polar groups are much more soluble in the sea water than the hydrocarbons themselves. In addition, clay minerals would absorb hydrocarbons and their derivatives and thus prevent such extensive accumulation at the surface of the oceans.

CARBON MONOXIDE

At high temperatures, the equilibrium in the reaction

$$CO + 3H_2 \leftrightarrows H_2O + CH_4$$

favors the formation of carbon monoxide. Consequently, volcanic gases that are "frozen" in a high-temperature equilibrium contain more CO than CH_4. This has led a number of workers to propose that CO was the main carbon species in the primitive atmosphere.

We think that even if carbon monoxide was the main carbon-containing molecule degassed from the primitive earth it could not have survived long in the atmosphere since it is thermodynamically unstable and relatively reactive. The reaction

$$CO + H_2 \rightarrow H_2CO$$

proceeds readily in ultraviolet light or in an electric discharge. This could have been the source of formaldehyde, a prebiotic precursor of sugars.

Carbon monoxide also reacts with water to give formate

$$CO + OH^- \rightarrow HCO_2^-$$

The half-life of this reaction would be about 12 million years at 0°C and 55,000 years at 25°C in oceans of contemporary size at pH 8.1.

When we take account of these and other reactions we conclude that CO could not have constituted the major carbon compound in the atmosphere. On the other hand, photochemical and volcanic production could have maintained a concentration much above the very low equilibrium concentration for a long time.

It should be noted that amino acids, and other organic compounds, can be formed from $CO—H_2—N_2$ mixtures. Even if we are wrong about the concentration of CO in the earth's primitive atmosphere, it would not invalidate the qualitative arguments to be given in the following chapters.

NITROGEN IN THE PRIMITIVE ATMOSPHERE

The stable form of nitrogen in the presence of hydrogen is NH_3. In discussing the history of the atmosphere-ocean-sediments system there are two further factors to be considered. Ammonia is very soluble in water and the dissolved ammonia would have been converted to NH_4^+ if the buffer capacity of the oceans and sediments was sufficient to maintain the pH at 8.1. The molar concentration of ammonia in water at 25°C is given by

$$\frac{p_{NH_3}}{NH_3} = 1.4 \times 10^{-2} \text{ atm M}^{-1}$$

Since the pK_a of NH_4^+ is 9.3 at 25°C, the ratio of NH_3/NH_4^+ is 0.06 at pH 80. Thus a 1 M NH_3 solution has a p_{NH_3} of 1.4×10^2 atm, while a 1 M NH_4Cl solution at pH 8.1 has a p_{NH_3} of 8.4×10^{-4} atm.

The second factor that affects the concentrations of NH_3 and NH_4^+ is the absorption of NH_4^+ into the clay minerals. The ammonium ion is quite similar to potassium ion in size, and it can exchange for the potassium in clays.

$$NH_4^+ + \text{Clay} \cdot K^+ \rightleftarrows K^+ + \text{Clay} \cdot NH_4^+$$

The equilibrium constant is about 1.0 for this reaction. The concentration of K^+ in the oceans is now 0.01 M, so that if the NH_4^+ concentration in the primitive ocean exceeded this figure, it seems likely that the NH_4^+ would have been taken up by clay minerals. Therefore, the primitive oceans were not likely to have contained more than 0.01 M NH_4^+.

The geological evidence alone does not give a lower limit for the concentration of ammonia in the primitive ocean. There could have been none at all, since there is none presently. However, an approximate lower limit can be obtained if we are prepared to assume that certain important prebiotic organic reactions, such as the synthesis of amino acids, took place in the primitive ocean. It should be clear to the reader that we are now using evidence that is not geophysical in origin.

A special example is the reversible deamination of aspartic acid to fumaric acid and NH_4^+.

$$^-OOC{-}CH_2{-}\underset{\displaystyle NH_3^+}{\underset{|}{CH}}{-}COO^- \rightleftarrows {}^-OOC{-}CH{=}CH{-}COO^- + NH_4^+$$

$$(NH_4^+) = K\,\frac{(\text{DL-aspartate})}{(\text{Fumarate})}$$

This reaction is quite rapid on the geological time scale ($t_{1/2} = 29$ million years at 0°C and 96,000 years at 25°C), so equilibrium would have been approximated in 10^7 to 10^8 years.

The value of K is 1.0×10^{-3} at 0°C and 2.7×10^{-3} at 24°C. If we assume that aspartic acid was needed for the evolution of life, and further assume that the aspartate/fumarate ratio did not fall substantially below 1.0, then the lower limit on the NH_4^+ concentration is about 1.0×10^{-3} M. An NH_4^+ concentration between 10^{-3} and 10^{-2} M seems reasonable for the prebiotic synthesis of most organic compounds of biological interest.

With these limits on the NH_4^+ concentration in the ocean, we can calculate limits on p_{NH_3} and p_{H_2} in the atmosphere if we assume that equilibrium was reached. The values given in Table 4-4 are based on an assumed p_{N_2} of one atmosphere.

Equilibrium between aqueous ammonia, aqueous ammonium ion, and atmospheric ammonia would undoubtedly have been attained. The reaction

Table 4-4. Equilibrium Concentrations of Ammonia and Hydrogen

	Temperature (°C)		
Item	0	25	50
	Upper limit from clay mineral equilibrium		
NH_4^+ (M)	0.01	0.01	0.01
NH_3 (M)	6.0×10^{-5}	4.2×10^{-4}	2.1×10^{-4}
p_{NH_3} (atm)	2.9×10^{-7}	7.3×10^{-6}	1.0×10^{-4}
p_{H_2} (atm)	1.6×10^{-7}	4.3×10^{-6}	6.3×10^{-5}
	Lower limit from aspartic acid equilibrium		
NH_4^+ (M)	1.0×10^{-3}	2.7×10^{-3}	5.8×10^{-3}
NH_3 (M)	6.0×10^{-6}	1.1×10^{-4}	1.2×10^{-3}
p_{NH_3} (atm)	2.9×10^{-8}	1.9×10^{-6}	5.9×10^{-5}
p_{H_2} (atm)	3.5×10^{-8}	1.8×10^{-6}	4.5×10^{-5}

of N_2 with H_2 would probably not have reached equilibrium because of the photodecomposition of ammonia. However, we think that the p_{NH_3} values given in Table 4-4 might have been maintained if p_{H_2} was 100 to 1,000 times the equilibrium value, that is, if p_{H_2} was between 10^{-5} and 10^{-3} atm. This value is not unreasonable. The table makes it clear that the atmosphere could not have contained large amounts of NH_3, because the absorption of NH_4^+ by clay minerals would have led to the removal of NH_3 from the atmosphere.

The argument given above depends on a number of assumptions, in particular the assumption that the accumulation of the aspartic acid needed by the first organisms occurred in the oceans or in water which had the same NH_4^+ content as the oceans. Since life may have evolved in lakes or tidepools, our conclusion about the ammonia content of the oceans is tentative.

This discussion makes it clear that the primitive atmosphere, often spoken of as a methane-ammonia atmosphere, must have been a methane-nitrogen atmosphere with traces of NH_3. Nonetheless, the ammonia in the oceans, in equilibrium with such an atmosphere, would have been sufficient to permit the synthesis and survival of organic compounds of biological interest.

OXYGEN IN THE ATMOSPHERE

The large amount of O_2 in our atmosphere is maintained by biological photosynthesis. The average O_2 production over the whole earth is 0.075 g cm^{-2} yr^{-1}; much of this is produced in the ocean (3.4×10^{17} g/yr). Almost all the O_2 production is used up by organisms oxidizing their food to CO_2. If this process did not occur, the amount of oxygen in the atmosphere (230 g cm^{-2}) would double in about 3,000 years.

In addition to the photosynthetic production of O_2 by plants, O_2 is produced by the ultraviolet photolysis of water in the upper atmosphere

$$2H_2O \rightarrow 2H_2 + O_2$$

The amount is quite small, about 10^{-8} g cm^{-2} yr^{-1}. In the absence of plants, however, significant traces of O_2 could be added to the atmosphere by this reaction. The steady state concentration maintained in the atmosphere in this way would be low since O_2 would react with reduced compounds (such as Fe^{++} and H_2S) on the surface of the earth.

Organic compounds are unstable in the presence of molecular oxygen (see Chapter 9). It is generally believed, therefore, that life arose and evolved to a considerable extent before a significant amount of O_2 appeared in the atmosphere from the photodissociation of H_2O.

Beyond the fact that there was no oxygen in the primitive atmosphere and that almost all oxygen is now produced by photosynthesis, there is little agreement in discussions of the course of development of oxygen in our atmosphere. The following arguments are usually held to be relevant.

(1) The Onverwacht and Fig-Tree organisms ($\sim 3.2 \times 10^9$ years). If these organisms are fossil blue-green algae, then they would have produced O_2 in the substantial quantities. However, the "substantial quantities" could have been as low as 0.1% of the present atmospheric level (0.2 atm) or as high as 100% of the present O_2 level (or even outside these bounds).

The Onverwacht and Fig-Tree organisms however may have been photosynthetic bacteria, which look very similar morphologically to blue-green algae. In that case, the atmosphere must have been anaerobic, since all photosynthetic bacteria are anaerobic.

To make matters more uncertain, the Onverwacht and Fig-Tree organisms may not be genuine microfossils or may have been formed in an atypical environment. Clearly no safe conclusions about the buildup of oxygen can be drawn from these micropaleontological data.

(2) Bulawayan limestone and gunflint organisms (2.7 to 3.0×10^9 years). The Bulawayan limestone was probably laid down by blue-green algae at this period, but again we cannot be sure of the level of O_2 that this would have produced. The presence of $CaCO_3$ implies that substantial amounts of CO_2 were available, but this does not imply any appreciable level of O_2. Indeed, substantial quantities of CO_2 would be expected even in a strongly reducing atmosphere. We can only conclude that some oxygen was probably present in the atmosphere 2.7×10^9 years ago, but it could have been only a trace (say 10^{-6} atm) or it could have been much more.

(3) Uranite and Galena deposits (2.0 to 2.8×10^9 years). The gold-uranium deposits of the Dominion Reef and Witwatersrand system in South Africa contain both Uranite (UO_2) and Galena (PbS). These minerals are easily oxidized to UO_3 and $PbSO_4$, respectively, at low partial pressures of O_2 ($\sim 10^{-20}$ atm O_2 at equilibrium). Therefore, the presence of these reduced compounds has been used as an argument for the virtual absence of O_2 in this period. However, these minerals may have been deposited under local reducing conditions, or had failed to have reached equilibrium with the atmosphere at the time they were laid down.

(4) Banded iron formations (1.8 to 2.2×10^9 years). The banded iron formations are enormous deposits of magnetite (Fe_3O_4 which can be considered an equimolar mixture of FeO and Fe_2O_3). The term *banded* refers to the alternating layers of iron rich and iron poor silica. It is held that these deposits were formed by the weathering of ferrous iron from various rocks. Ferrous iron is relatively soluble and so would have been transported by streams. Molecular oxygen in the atmosphere would then have oxidized the ferrous iron to ferric iron, which would have precipitated (with some of the

remaining ferrous iron) as Fe_3O_4. This mechanism implies that substantial amounts of O_2 were present in the atmosphere. Again, it is not clear how high the level would have been. It is also possible that the ferrous iron was oxidized, perhaps photosynthetically by bacteria in the absence of oxygen in the atmosphere, or by a nonbiological process.

This picture is complicated by the presence of red beds, which are deposits of Fe_2O_3 (hematite). It probably requires higher oxygen pressures to produce Fe_2O_3 than the Fe_3O_4 of the banded iron formations. Most of the red beds were laid down later than the banded iron formations, but some are younger. These observations can be accounted for either by supposing that local areas of high oxygen pressure existed or by assuming that magnetite was oxidized to hematite many years after its deposition.

(5) Appearance of eukaryotic organisms ($\sim 1.2 \times 10^9$ years). Eukaryotic organisms are aerobic, so the appearance of these organisms may imply that large amounts of O_2 were found in the atmosphere. On the other hand, it may not. Modern eukaryotic organisms may have adapted to higher levels of O_2 than were encountered by the first eukaryotes.

(6) The Pasteur effect. Facultative aerobes (e.g., E. coli and yeasts) will ferment glucose when the oxygen pressure is low, but switch over to oxidative metabolism when the oxygen pressure is high. The change takes place in many microorganisms at 1% of the present O_2 level. We do not see that this can be used reliably to estimate the O_2 pressure in the primitive atmosphere at any particular time.

(7) Explosive evolution at the base of the Cambrian period ($\sim 0.6 \times 10^9$ years). The striking development of many species of multicellular organisms over a short period at the beginning of the Cambrian argues in favor of the presence of important quantities of O_2 in the atmosphere. It is very unlikely that large multicellular organisms could have existed on a fermentative type of metabolism (2 molecules of ATP are obtained by fermentation of glucose to lactic acid, but 36 molecules of ATP are obtained by oxidation of glucose to CO_2 and water). We may, therefore, feel confident that at least several percent of the present O_2 level had accumulated in the atmosphere by the beginning of the Cambrian. However, we cannot attribute the explosive evolution of higher species to a sudden buildup of O_2, since many other explanations have been proposed.

These factors have been put together by various authors in different ways to estimate the time course of O_2 in the atmosphere. In Figure 4-4, we show the estimates given in Berkner and Marshall and by Rutten. Also shown are what we consider a high O_2 model and a low O_2 model. There is clearly a wide range of choices of intermediate models. These models have all assumed a monotonically increasing concentration of O_2 in the atmosphere. It is clear that fluctuating O_2 levels are also possible and that some of the contradictory points could be resolved on this basis. The reader may wish

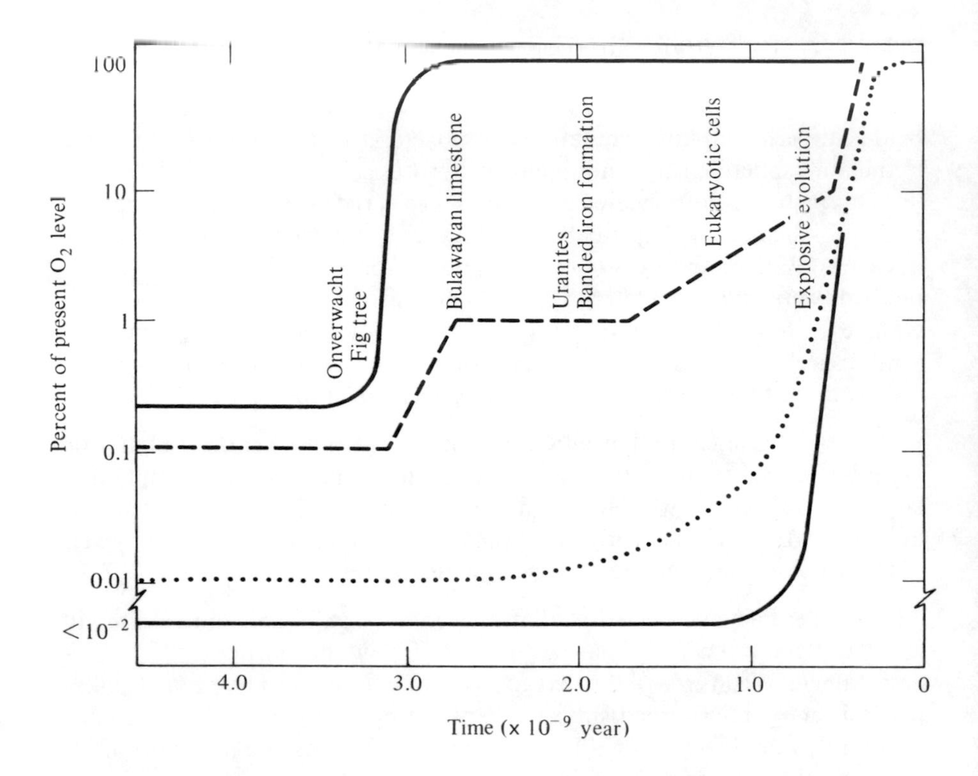

Fig. 4-4. Estimates of the oxygen levels in the earth's atmosphere at different periods. The upper solid line is a high estimate of the O_2 partial pressure. The lower solid line is a low estimate. The dotted line is the estimate by Berkner and Marshall and the dashed line is the estimate by Rutten. The maximum percent of O_2 is shown at 100%, but it may have been higher than this in the past. The reader can easily construct alternative models, which may include fluctuations in the O_2 partial pressure.

to construct his own scenario for the time course of development of O_2 in the atmosphere.

References

P. J. Brancazio and A. G. W. Cameron, *The Origin and Evolution of Atmospheres and Oceans* (John Wiley, New York, 1964).

H. C. Urey, "On the early chemical history of the earth and the origin of life," *Proc. Nat. Acad. Sci.* **38**, 349 (1952).

H. C. Urey, "Regarding the early history of the earth's atmosphere," *Bull. Geol. Soc. Amer.* **67**, 1125 (1956).

W. W. Rubey, "Development of the hydrosphere and atmosphere, with special reference to probable composition of the early atmosphere," in *Crust of the Earth*, A. Poldervaart, ed. (Geol. Soc. Am., New York, 1955), pp. 631–650. *Bull. Geol. Soc. Am.* **62**, 1111 (1951).

H. D. Holland, "Model for the evolution of the earth's atmosphere," in *Petrologic Studies: A Volume to Honor A. F. Buddington* (Geological Society of America, New York, 1962), pp. 447–477.

H. D. Holland, "The history of ocean water and its effect on the chemistry of the atmosphere," *Proc. Nat. Acad. Sci.* **53**, 1173 (1965).

H. D. Holland, "The geologic history of sea water—an attempt to solve the problem," *Geochim. Cosmochim. Acta* **36**, 637 (1972).

A. C. Lasaga, H. D. Holland, and M. J. Dwyer, "Primordial oil slick," *Science* **174**, 53 (1971).

L. G. Sillén, "Oxidation state of earth's ocean and atmosphere. I. A model calculation on earlier states. The myth of the 'prebiotic soup'," *Arkiv Kemi* **24**, 431 (1965). II. The behavior of Fe, S and Mn in earlier states. Regulating mechanisms for O_2 and N_2, *Arkiv Kemi* **25**, 159 (1965).

L. G. Sillén, "Regulation of O_2, N_2 and CO_2 in the atmosphere, thoughts of a laboratory chemist," *Tellus* **18**, 198 (1966).

L. G. Sillén, "The ocean as a chemical system," *Science* **156**, 1189 (1967).

F. T. MacKenzie and R. M. Garrels, "Chemical mass balance between rivers and oceans," *Am. J. Sci.* **264**, 507 (1966).

S. I. Rasool and W. E. McGovern, "Primitive atmosphere of the earth," *Nature* **212**, 1225 (1966).

P. H. Abelson, "Chemical events on the primitive earth," *Proc. Nat. Acad. Sci. U.S.* **55**, 1365 (1966).

J. E. Van Trump and S. L. Miller, "Carbon monoxide on the primitive earth," *Earth Planet. Sci. Lett.* (in press, 1973).

J. L. Bada and S. L. Miller, "Ammonium ion concentration in the primitive ocean," *Science* **159**, 423 (1968).

R. A. Raff and G. M. Meaburn, "Photochemical reaction mechanisms for production of organic compounds in a primitive earth atmosphere," *Nature* **221**, 459 (1969).

L. V. Berkner and L. C. Marshall, "On the origin and rise of oxygen concentration in the earth's atmosphere," *J. Atmos. Sci.* **22**, 225 (1965).

R. T. Brinkmann, "Dissociation of water vapor and evolution of oxygen in the terrestrial atmosphere," *J. Geophys. Res.* **74**, 5355 (1969).

P. E. Cloud, Jr., "Atmospheric and hydrospheric evolution on the primitive earth," *Science* **160**, 729 (1968).

M. G. Rutten, *The Origin of Life* (Elsevier, Amsterdam, 1971).

M. G. Rutten, "The history of atmospheric oxygen," *Space Life Sciences* **1**, 1–13 (1970).

L. Van Valen, "The history and stability of atmospheric oxygen," *Science* **171**, 439 (1971).

L. Sagan, "On the origin of mitosing cells," *J. Theoret. Biol.* **14**, 225 (1967).

L. Margulis (Sagan), *Origin of Eukaryotic Cells* (Yale University Press, New Haven, Conn., 1970).

Chapter Five

Sources of Energy

An atmosphere of CH_4, NH_3, N_2, H_2O, and H_2 in equilibrium at a moderate temperature, or any other reasonable atmosphere at equilibrium, does not contain or deposit significant quantities of organic compounds. However, in the presence of a source of energy, organic compounds can be synthesized in a reducing atmosphere. The absorption of energy in the atmosphere leads to the synthesis of reactive intermediates, which then combine to form more complex organic compounds.

A number of important points are illustrated by the following scheme for the synthesis of glycine.

$$2CH_4(g) + NH_3(g) + 2H_2O(l) \rightleftarrows H_2N{-}CH_2{-}COOH\,(aq) + 5H_2(g) \qquad \Delta G° = +52.2 \text{ kcal}$$

$$2CO(g) + NH_3(g) + H_2(g) \rightleftarrows H_2N{-}CH_2{-}COOH\,(aq) \qquad \Delta G° = -19.5 \text{ kcal}$$

$$H_2CO(g) + HCN(g) + NH_3(g) + H_2O(l) \rightleftarrows H_2N{-}CH_2{-}COOH\,(aq) + NH_3(g) \qquad \Delta G° = -91.6 \text{ kcal}$$

$$2CH_4(g) + NH_3(g) + \tfrac{5}{2}O_2(g) \rightleftarrows H_2N{-}CH_2{-}COOH\,(aq) + 3H_2O(l) \qquad \Delta G° = -230.9 \text{ kcal}$$

The free energy of formation of glycine from CH_4 and NH_3 is so unfavorable that the reaction is impossible. Synthesis from CO and NH_3 is possible, but not particularly favorable (-19.5 kcal is the free energy under standard conditions of 1 atm and 1 M aqueous solutions). The synthesis of glycine from formaldehyde, ammonia and hydrogen cyanide is accompanied by a much greater release of free energy. This reaction occurs readily although the synthesis from CO and NH_3 does not at 25°C. The free energy is even greater for the fourth reaction, but this reaction does not take place. Thus a favorable free energy is a necessary, but not a sufficient condition for a reaction to occur. Within the limits imposed by thermodynamics, kinetics dominates organic chemistry and particularly the organic chemistry which interests us most.

There would, of course, be no accumulation of organic compounds if (once synthesized) they reacted rapidly to reform an equilibrium mixture (e.g., of CH_4, NH_3, H_2O, and H_2). Thus, the accumulation of organic compounds on the primitive earth must have depended on the synthesis of "high-energy," reactive, simple organic compounds that combined to form more complex organic compounds. These more complex organic compounds, while thermodynamically unstable, must have been kinetically stable so that they could accumulate in the primitive ocean or lakes. Some of the organic compounds of biological interest are kinetically stable for several billion years. Other important compounds, for example sugars, would be stable only for short periods of time; the accumulation of these latter compounds presents special problems (Chapter 9).

The direct sources of energy available for organic synthesis on the primitive earth are sunlight, electric discharges, thermal energy of volcanoes, cosmic

Table 5-1. Present Sources of Energy Averaged Over the Earth[a]

Source	Energy (cal cm^{-2} yr^{-1})
Total radiation from sun	260,000
Ultraviolet light	
< 3,000 Å	3,400
< 2,500 Å	563
< 2,000 Å	41
< 1,500 Å	1.7
Electric discharges	4
Cosmic rays	0.0015
Radioactivity (to 1.0 km depth)	0.8
Volcanoes	0.13
Shock waves	1.1

[a] The figure for electric discharges includes 0.9 cal cm^{-2} yr^{-1} from lightning and about 3 cal cm^{-2} yr^{-1} from corona discharges from pointed objects [B. Schonland, *Atmospheric Electricity* (Methuen, London, 1953), pp. 42, 63]. The value for radioactivity 4×10^9 years ago was 2.8 cal cm^{-2} yr^{-1} [E. Bullard, in *The Earth as a Planet*, G. P. Kuiper, ed. (University of Chicago Press, Chicago, 1954), p. 110]. The volcanic energy is based on the emission of 1 km^3 of lava per year at 1000°C ($Cp = 0.25$ cal gm^{-1}, $\rho = 3.0$ cal gm^{-1}). The energy from shock waves includes 0.1 cal cm^{-2} yr^{-1} from meteorites and meteors and 1 cal cm^{-2} yr^{-1} from the shock or pressure wave of a lightning bolt [A. Bar-Nun *et al.*, *Science* **168**, 470 (1970)].

rays, radioactivity, and shock waves. The amount of energy available from these sources is given in Table 5-1.

The largest source of energy is clearly the sun, but only a small portion of its output is at wavelengths below 2,000 Å, where it can be absorbed by such compounds as CH_4, H_2O, and NH_3. Electric discharges are next in importance as a source of energy. The figure of 4 cal cm^{-2} yr^{-1} includes 0.9 cal from lightning and about 3 cal cm^{-2} yr^{-1} due to corona discharges from pointed objects.

The energy from cosmic rays is negligible at present, and we doubt that it was ever very important. Radioactive disintegration of uranium, thorium, and potassium would have provided 2.8 cal cm^{-2} yr^{-1}. Radioactivity is a poor source of energy for organic synthesis, since the U, Th, and K are

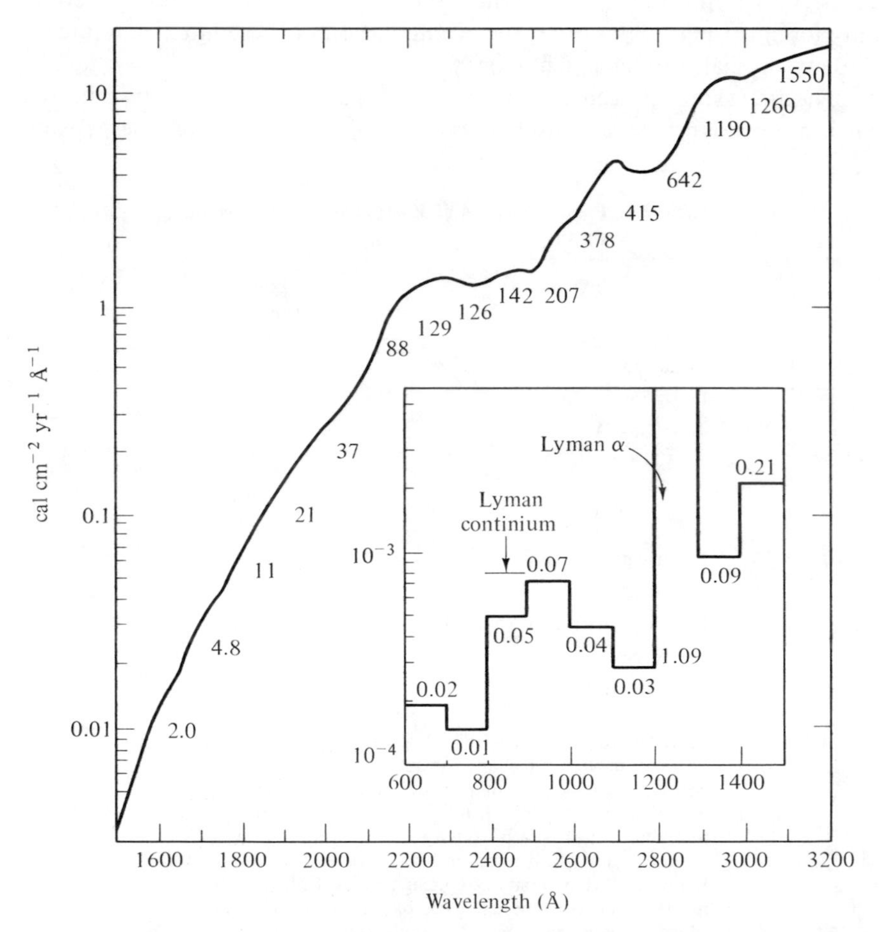

Fig. 5-1. Solar energy incident on the upper atmosphere, averaged over the surface of the earth, as a function of wavelength. The numbers beneath the curves are the energies available in 100 Å intervals—for example, 129 cal cm^{-2} yr^{-1} are available in the interval 2200–2300 Å.

concentrated in rocks, and so the energy they supply could not easily be used to break up or synthesize carbon compounds.

The energy available from volcanoes is rather small: Approximately 1 km^3 of lava is emitted per year at the present time. Even if it was an order of magnitude greater on the primitive earth, it would not make much difference, since volcanic energy is not effective in synthesizing organic compounds. Thermal energy can be used to synthesize organic compounds only if the constituents of the atmosphere are heated to 600°C or higher, and the products are then quenched rapidly. Therefore, only the surface of molten lava would be effective; once the surface of a lava flow became solid, the surface temperature would be nearly that of the atmosphere, even though the lava below the surface was molten. Thermal energy from volcanoes should not be dismissed entirely, but it could only have made a small contribution to organic compound synthesis on the primitive earth.

Other sources of energy have been proposed. Examples are shock waves from collisions of comets or meteorites with the atmosphere, sonic energy from ocean waves, cosmic or solar X rays, and the surface energy of bubbles formed on the surface of the ocean. We believe that in most cases, either the energies available or the conversion factors were too small for these sources of energy to have played a major role in prebiotic synthesis. Some surprisingly efficient syntheses of amino acids by shock waves have been reported. If

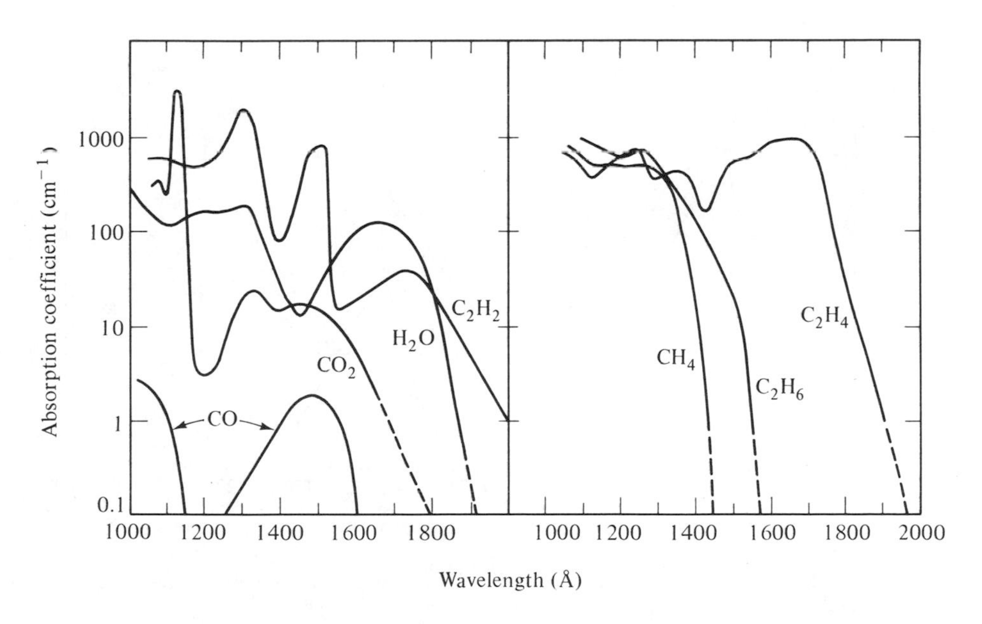

Fig. 5-2. Absorption spectra of CO, CO_2, H_2O, C_2H_2, CH_4, C_2H_6, and C_2H_4. The fine structure of these spectra has been smoothed out. An absorption coefficient of 1.0 means that 10% of the light is transmitted (90% absorbed) by a 1.0 cm column of the gas at 1 atm pressure and 0°C (optical density = 1).

amino acids are synthesized in 30–40% yield as claimed, this energy source could have been of considerable importance.

In the present atmosphere, ultraviolet light of wavelengths below about 3,000 Å does not reach the surface of the earth. Wavelengths below 2,200 Å are absorbed by O_2, and ozone absorbs the light between 2,200 and 3,200 Å. A reducing atmosphere would have been more transparent.

The wavelengths at which some of the constituents of a reducing atmosphere begin to absorb are shown in Figs. 5-2 and 5-3. All the components of a reducing atmosphere are transparent above 2,000 Å, except ammonia which has weak bands at about 2,250 Å and H_2S which absorbs out to about 2,800 Å. The amount of energy available from ultraviolet light for synthesis in a reducing atmosphere is thus between 41 cal (below 2,000 Å) and 1.7 cal (below 1,500 Å) (Table 5-1). NH_3 and H_2S could absorb larger amounts of energy, and this energy may have been important in prebiotic synthesis. The 1.7 cal cm^{-2} yr^{-1} available below 1,500 Å is made up of 0.7 cal from black body radiation, and 1.0 cal from the strong Lyman α line of hydrogen at 1,216 Å. If small amounts of more complex molecules (e.g., formaldehyde, acetone, benzene) were present in the atmosphere, longer wavelengths could be absorbed and much larger amounts of energy would have been available.

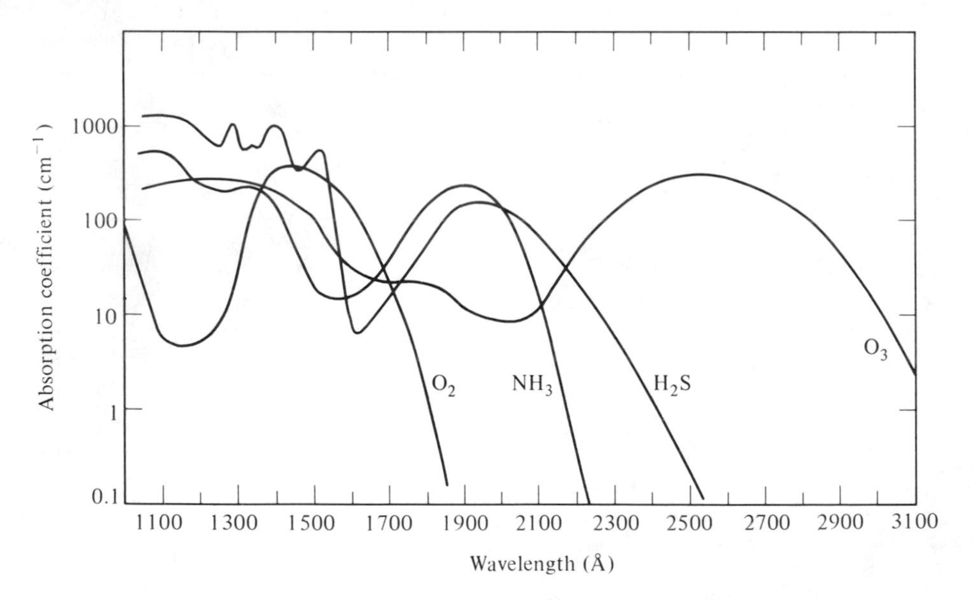

Fig. 5-3. Absorption spectra of O_2, NH_3, H_2S, and O_3. The fine structure of these spectra has been smoothed out. A gas with an absorption coefficient of 1.0 absorbs 90% of the light incident on the upper atmosphere when the partial pressure of the gas is approximately 10^{-6} atm. Under these conditions only 10% of the light reaches the surface of the earth.

Large amounts of energy are available in the form of visible and near-ultraviolet light. They would be ineffective in the synthesis of organic compounds in the atmosphere, but might have been important in the oceans, since many photochemical transformations in solution do not require large amounts of energy. An example would be the synthesis of cyanogen from ferricyanide.

The present day ocean is not very transparent to ultraviolet light owing to the presence of dissolved and colloidal iron compounds as well as some absorbing organic compounds. The primitive ocean was probably even less transparent. A yellow tar is produced by the action of electric discharges on CH_4 and NH_3, and a brown-black polymer is produced from cyanide solutions (Chapter 7). These and other organic materials absorb strongly in the ultraviolet. If such material were produced on the primitive earth, as is likely, ultraviolet light could have penetrated to only a very small depth in the ocean. Any photochemical reaction involving ultraviolet light would have taken place near the surface of the ocean. Organic compounds in the ocean would have been protected from destruction by the ultraviolet light, except when they came close to the surface.

The energy in lightning and corona discharges is comparable in amount to the ultraviolet light below 1,500 Å. Although lightning is more spectacular, the corona discharge from pointed objects is more abundant. It is likely that the corona discharges are more efficient in synthesizing organic compounds than lightning, since much of the energy of the lightning is converted into heat rather than used in the ionization of molecules.

At the present time, 4 cal cm^{-2} yr^{-1} of electrical energy is available; the value on the primitive earth is unknown, but it could have been substantially different. There are several mechanisms for charge generation in thunderstorms and the relative importance of these could be greatly altered in a reducing atmosphere, especially due to the presence of NH_3. Thunderstorm electricity is a complicated and poorly understood subject, and no estimates of the energy available from this source on the primitive earth have been made.

Electric discharges are very efficient in synthesizing hydrogen cyanide from CH_4 and NH_3 or N_2, but ultraviolet light seems to give poor yields. Although ultraviolet light was surely of great importance as a source of energy, it seems likely at the moment that electric discharges were the most important source of hydrogen cyanide and hence of those prebiological organic compounds formed from it. Later we shall show that hydrogen cyanide was an important intermediate in prebiotic syntheses (Chapter 7).

Electric discharges have another advantage over ultraviolet light: They would have occurred near the surface of the earth while most of the photochemical reactions brought about by ultraviolet light would have occurred in the upper atmosphere. If the compounds formed in this way absorbed light of longer wavelengths they might not have reached the oceans

Fig. 5-4. A lightning flash is the most dramatic form of atmospheric electricity. There are about 16×10^6 thunder storms per year on the earth and about 100 lightning flashes per second. A lightning flash averages 1.6×10^9 cal. If it takes 100 kcal of electrical energy to produce 1 mole of HCN, then the average flash would have produced about $\frac{1}{2}$ (metric) ton of HCN on the primitive earth. (From *Harper's Encyclopedia of Science*, Vol. 4, p. 1187.)

before they were photodissociated. The problem of the rate of transport to the oceans is a complicated one, so that it is not clear what fraction of the organic compounds formed in the upper atmosphere would reach the oceans before undergoing photodecomposition.

We can make some sample calculations of the yields of organic compounds from the energies available. If we assume that one mole of HCN was produced from 100 kcal of electrical energy, then 4×10^{-5} moles $cm^{-2}\ yr^{-1}$ HCN would have been produced. If this were dissolved in an ocean of the present size, the HCN concentration would increase by 1.3×10^{-7} M yr^{-1}. If we assume that aldehydes were produced from the Lyman α line (1.0 cal $cm^{-2}\ yr^{-1}$) with a quantum yield of 0.5, then 2×10^{-6} moles $cm^{-2}\ yr^{-1}$ would have been produced from this source. If this could have been transported to an ocean of the present size, the increase in the concentration would have been 0.7×10^{-8} M yr^{-1}. If the energy quantum yields, or efficiency of transport to the ocean were less than those assumed, the concentrations would have been less. On the other hand, if the oceans were smaller, the concentrations would have been correspondingly greater.

Compounds which do not hydrolyze or decompose rapidly can accumulate. Thus, a thousand years' production would give a HCN concentration of 1.3×10^{-4} M and an aldehyde concentration of 0.7×10^{-5} M with the above assumptions. The total production of HCN in 10^6 years would correspond to 0.13 M. In fact, the concentration could never approach this value because the HCN would hydrolyze or react in much shorter times (Chapter 7). On the other hand, the amino acids are relatively stable. If the yield of amino acids was equal to the production of aldehydes, 7×10^{-3} M amino acids would accumulate in a million years. If all the HCN went into the synthesis of adenine, the concentration would be 0.026 M after a million years. This is greater than its solubility. Even if we have been too optimistic in our estimates of quantum yields it is clear that very large amounts of organic material could accumulate in geologically reasonable times.

References

Table 5-1 is revised from a table in S. L. Miller and H. C. Urey, "Organic Compound Synthesis on the Primitive Earth," *Science* **130**, 245 (1959). It is based on new values for the ultraviolet flux from the sun [R. Tousey, *Space Science Reviews* **2**, 3 (1963); S. R. Pottasch, *Space Science Reviews* **3**, 816 (1964); H. E. Hinteregger, *Space Science Reviews* **4**, 461 (1965)].

The absorption spectra in Figs. 5-2 and 5-3 are taken mostly from K. Watanabe, "Ultraviolet Absorption Processes in the Upper Atmosphere," *Adv. Geophys.* **5**, 153 (1958). Other data are from R. E. Huffman *et al.*, *J. Chem. Phys.* **39**, 910 (1963) (N_2); T. Nakayama and K. Watanabe, *J. Chem. Phys.* **40**, 558 (1964) (C_2H_2); K. Watanabe and A. S. Jursa, *J. Chem. Phys.* **41**, 1650 (1964) (H_2O and H_2S); B. A. Thompson *et al.*, *J. Geophys. Res.* **68**, 6431 (1963).

Atmosphere electricity is poorly understood, especially the mechanisms of charge generation. An elementary book on the subject is B. F. J. Schonland, *Atmospheric Electricity* (Methuen, London, 1953). More advanced are H. R. Byers, *Thunderstorm Electricity* (University of Chicago Press, Chicago, 1953), and M. A. Uman, *Lightning* (McGraw-Hill, New York, 1969).

Discussions of the role of shock waves and sonic energy from cavitation in liquid water are given in A. R. Hochstim, *Proc. Nat. Acad. Sci. USA* **50**, 200 (1963); A. Bar-Nun, N. Bar-Nun, S. H. Bauer, and C. Sagan, *Science* **168**, 470 (1970), and M. Anbar, *Science* **161**, 1343 (1968).

Chapter Six

Nature of Biology

INTRODUCTION

It is not possible to give a coherent discussion of theories on the origin of life without taking account of the basic processes of biology. Many of the attributes of particular organisms are the products of later evolution and will not be of great interest to us. The processes of nucleic acid replication and protein synthesis are universal in biology, so some of these processes were probably involved in the origin of life.

THE ENZYMES AND OTHER PROTEINS

It is helpful to think of a cell as a factory in which some hundreds or thousands of elementary chemical operations are smoothly coordinated. In this factory the proteins play a multiple role as machines, control elements, and sometimes also as structural components. We shall be concerned mainly with the enzymes; that is, the machine-like proteins which are responsible for the efficient catalysis of all the varied reactions needed for cellular function.

A protein consists of one or a small number of polypeptide chains. Each of these is a polymer made up of a number of different α-amino acids having the L-optical configuration, joined together as illustrated in Fig. 6-1. It is a remarkable experimental finding that in all living things the same set of 20 α-amino acids (Fig. 6-2) are utilized in building proteins. The few unusual amino acids found in special proteins—for example, hydroxy proline in collagen—have turned out not to be exceptions to this rule because they

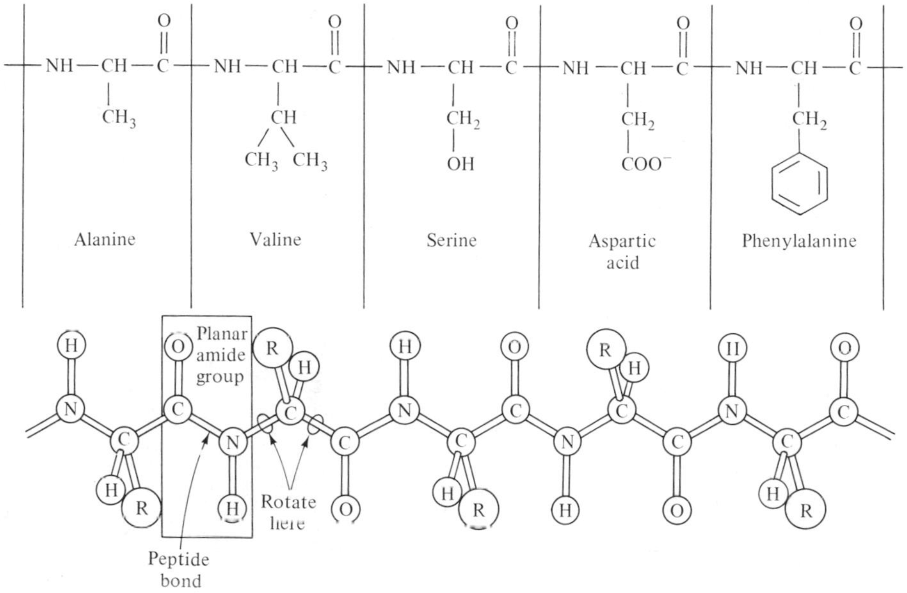

Fig. 6-1. A portion of a polypeptide chain. The upper diagram shows a chemical formula. The lower diagram shows one structural representation—a zigzag backbone with the side chains on alternate sides of the backbone.

are made by modification of one of the standard 20 amino acids subsequent to the synthesis of the polypeptides. Antibiotics often contain unusual amino acids, for example, ornithine and D-phenylalanine in gramacidins. These antibiotics are not produced by the normal mechanism of protein synthesis, but are made by a set of special enzymes.

The identity of a protein is determined in the first place by the sequence of amino acids in its polypeptide chains. This sequence defines the primary structure of the protein. In Fig. 6-3 we illustrate the primary structure of a typical protein, the hydrolytic enzyme, lysozyme. The polypeptide chains making up enzymes usually contain at most a few hundred amino acids; a few proteins, for example insulin, have much lower molecular weights.

The efficiency of an enzyme depends on the accurate positioning in space of the functional groups responsible for catalysis; these latter are said to constitute the *active site*. The specificity of enzymes—for example, the ability to distinguish between such closely related substrates as valine and isoleucine (Fig. 6-2)—is thought to depend on a group of amino acid side chains which constitute the *specificity site*. The specificity site should not be thought of as necessarily distinct from the active site. Many enzymes include in addition to the active site and specificity site special regions called *allosteric sites* which permit the enzyme function to be controlled by molecules other than

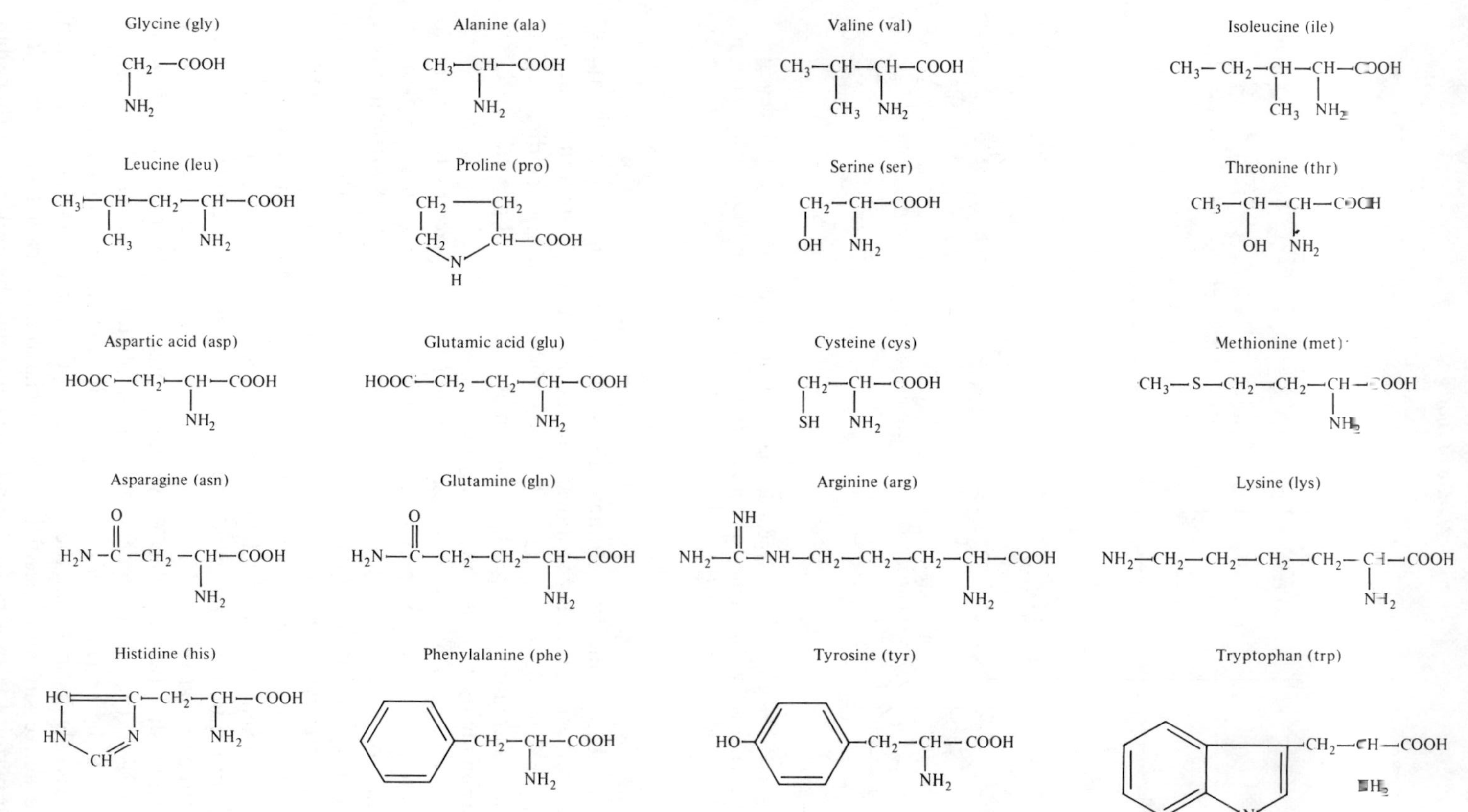

Fig. 6-2. The 20 amino acids found in proteins.

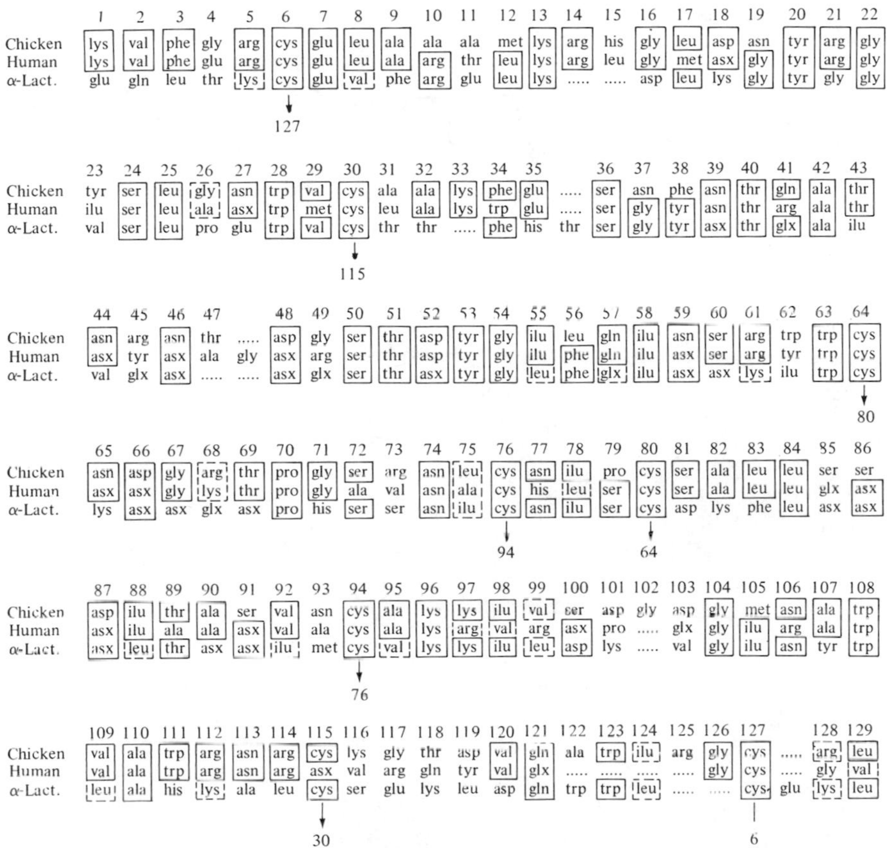

Fig. 6-3. The primary sequences of chicken egg white lysozome and human lysozome. The structure of α-lactoglobulin, a protein with a completely different function, is included in the diagram. Horizontal dots indicate a gap, either from a deletion in one sequence or an insertion in another. Identical residues in two or three of the sequences are outlined with a solid line. Very similar residues (e.g., lys for arg) are outlined with a dotted line. Asx and glx mean that it is uncertain whether the residue is aspartic acid or asparagine and glutamic acid or glutamine. The most favorable choice is assumed for asx and glx in drawing the homologies. The numbers beneath the cysteine residues indicate the partners in the four disulfide bridges. It is clear that lysozine and α-lactoglobulin have evolved from a common ancestral protein.

the substrate. This allows sequences of reactions to be integrated into metabolic pathways.

For an enzyme to be functional, it is not only necessary that the primary sequence be correct, but also that the three-dimensional structure of the polypeptide chains be maintained. The denaturation of a protein involves the modification of the three-dimensional structure, usually without change in the primary sequence. The forces maintaining the structure of a protein are of various types—for example, hydrophobic interaction between hydrocarbon side chains, electrostatic forces between acidic and basic residues,

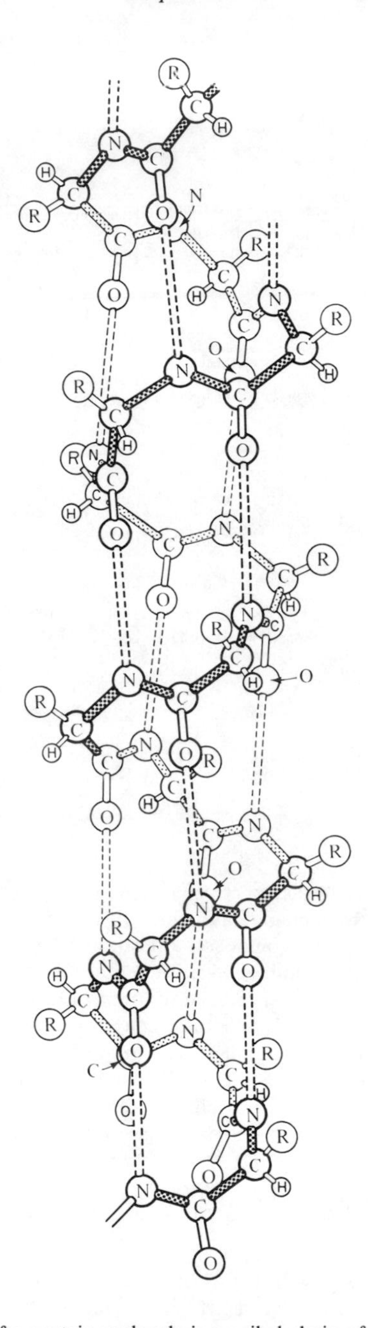

Fig. 6-4. *Alpha helix* of a protein molecule is a coiled chain of amino acid units. The backbones of the units form a repeating sequence of atoms in carbon (C), oxygen (O), hydrogen (H), and nitrogen (N). The R stands for the side chain that distinguishes one amino acid from another. The configuration of the helix is maintained by hydrogen bonds (broken lines). The hydrogen atom that participates in each of these bonds is not shown. (From Kendrew, *Scientific American*, December, 1961.)

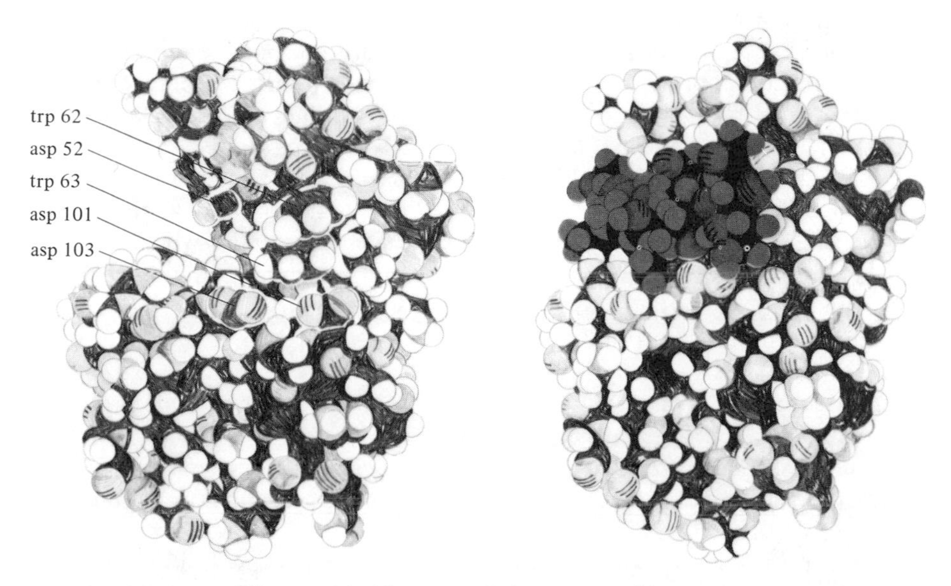

Fig. 6-5. Space-filling model of lysozome. Left: Enzyme without substrate showing active site crevice. Right: Enzyme-substrate complex.

and hydrogen bonding. In special cases S-S bonds and side-chain interactions with metals are also important. In polypeptides made up from a single aliphatic amino acid, the structure is often dominated by hydrogen bonding between the C=O and NH groups of the peptide chain. The structure of these simple homopolypeptides is usually the α helix, illustrated in Fig. 6-4. In proteins, the amount of α helix is variable, but rarely exceeds 50% of the residues. This is illustrated in Fig. 6-5, which shows the secondary and tertiary structure of lysozyme (cf. Fig. 6-3).

The great superiority of enzymes over other homogeneous catalysts is due in part to their complex but precisely defined structures. An enormous number of such structures can be built using sufficiently long strings of the 20 amino acids. By utilizing residues which are far apart along the polypeptide chain to make a "cavity" around the substrate molecule (Fig. 6-5), a specificity is achieved which would hardly be possible using the small molecules usually made by synthetic organic chemists.

Inspection of Fig. 6-2 shows that the amino acid side chains are of a number of types. Firstly, there are the "nonfunctional" amino acids—glycine, alanine, valine, leucine, isoleucine, phenylalanine, proline, and perhaps tryptophan—all of which must be important in determining secondary structure and in forming specificity sites, but which are unlikely to function directly in catalysis. Next there are the hydroxyl-containing amino acids—serine, threonine, and tyrosine. Serine in particular has been implicated in the active sites of many enzymes. In hydrolytic enzymes, it often reacts with the substrate to form an ester which is subsequently broken down to complete the reaction.

Another important group of amino acids are the polar amino acids—aspartic acid, glutamic acid, lysine, arginine, and histidine. These residues play an important role in determining the solubility and polyelectrolyte properties of proteins. In addition, histidine is believed to be a particularly important catalytic group, while lysine has been shown to be the combining site of certain decarboxylating enzymes. There is no doubt that the other polar amino acids also function at the active site in certain enzymes.

No clear catalytic activity has been established for asparagine and glutamine; they are likely to be important in recognizing polar molecules, such as purines and pyrimidines. The role of methionine in enzymes is also unclear, but the other sulphur-containing amino acid, cysteine, is extremely important for two reasons. Firstly, the secondary structure of some proteins is maintained by S-S links formed by the oxidation of pairs of cysteine residues. Secondly, free SH groups play a major role in several active sites, sometimes as directly reacting groups and sometimes as attachment sites for heavy metals.

When we begin to consider problems of biochemical evolution, we shall be particularly concerned to know how much of a protein must be specified in order to obtain catalytic activity and how much is either capable of variation or is involved in secondary functions necessary for control and integration within the cell. This question cannot be answered at present, but we do have some relevant information.

Comparison of pairs of enzymes with similar function, isolated from different organisms, often shows that although they have a common origin they have been so modified by mutation that only a small proportion of the residues are the same in the two structures. Thus, at most positions in the polypeptide chain of an enzyme, the amino acid can be changed to a similar amino acid without affecting the type of enzyme activity. In other words, there are very many amino acid sequences which have the same enzyme activity.

Crystallographic studies suggest that the active site involves a small number of functional groups, a reasonable guess might be two or three. Some further residues must be needed to define the specificity site, and many more to guarantee the correct secondary structure. However, an enzyme with a reasonable activity could probably be specified using much less information than is required to determine the primary structure of a contemporary protein. Thus, calculations which suggest that the probability of creating an enzyme of 100 amino acids by chance is 20^{-100} are fallacious.

NUCLEIC ACIDS

Nucleic acids are mainly involved in the determination of the primary sequence of proteins. We shall see that the genetic material, deoxyribonucleic

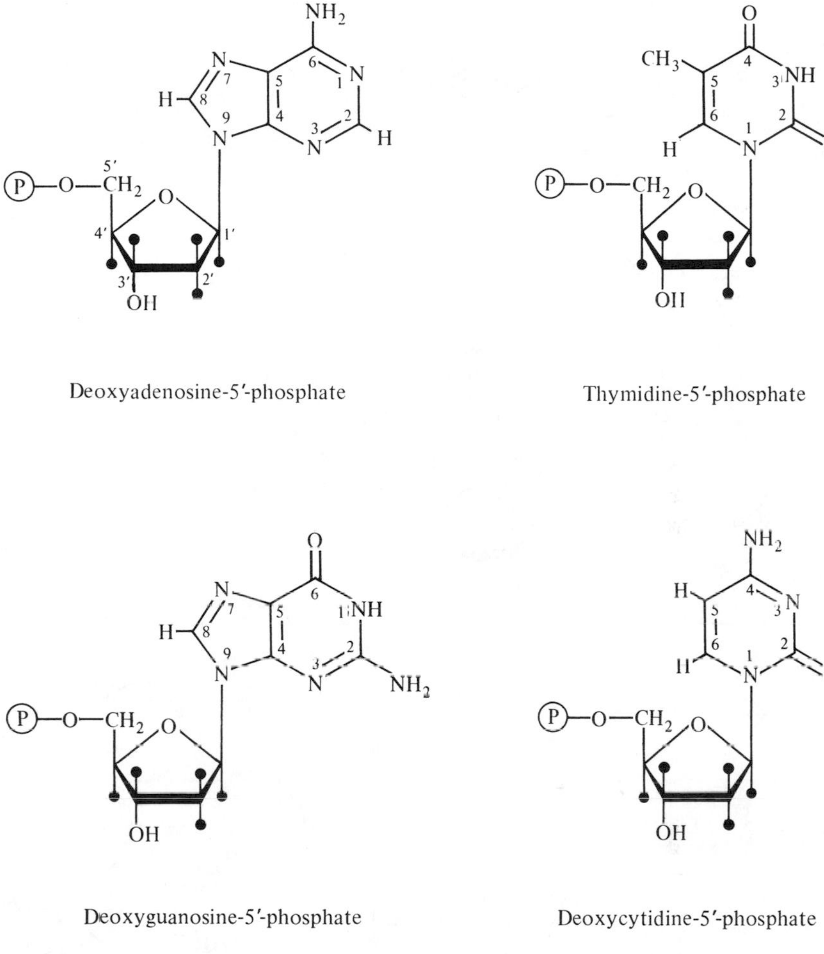

Deoxyadenosine-5′-phosphate

Thymidine-5′-phosphate

Deoxyguanosine-5′-phosphate

Deoxycytidine-5′-phosphate

Fig. 6-6. The four deoxynucleotides occurring in DNA.

acid (DNA) and certain types of ribonucleic acid (RNA), are information stores for protein synthesis; some RNA is involved in protein synthesis but does not carry genetic information (transfer RNA and ribosomal RNA).

The genetic material of all organisms other than a few viruses is two-stranded DNA. This polymeric molecule is made up from four monomeric components known as *deoxyribonucleotides.* They are deoxyadenylic acid (d-A), deoxyguanylic acid (d-G), thymidylic acid (T)* and deoxycytidylic acid (d-C). The structure of these molecules and the numbering system is indicated in Fig. 6-6. In some viruses that contain two-stranded DNA as

* The standard notation is inconsistent since thymidylic acid contains deoxyribose. Logic would require d-T. The riboside of thymine is called 5-methyl uridine.

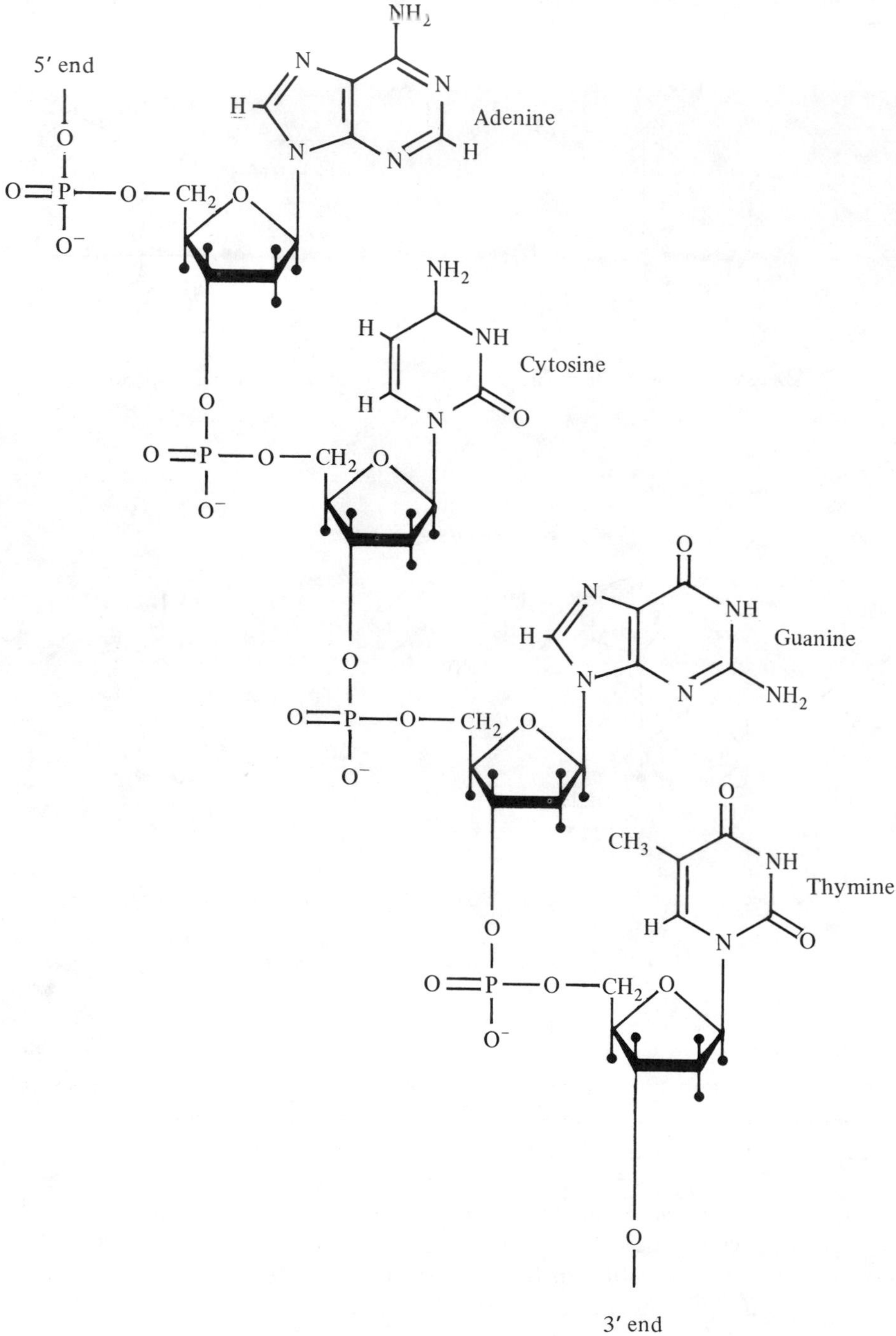

Fig. 6-7. A section from a DNA chain showing the sequence ACGT. (From Watson, 1970.)

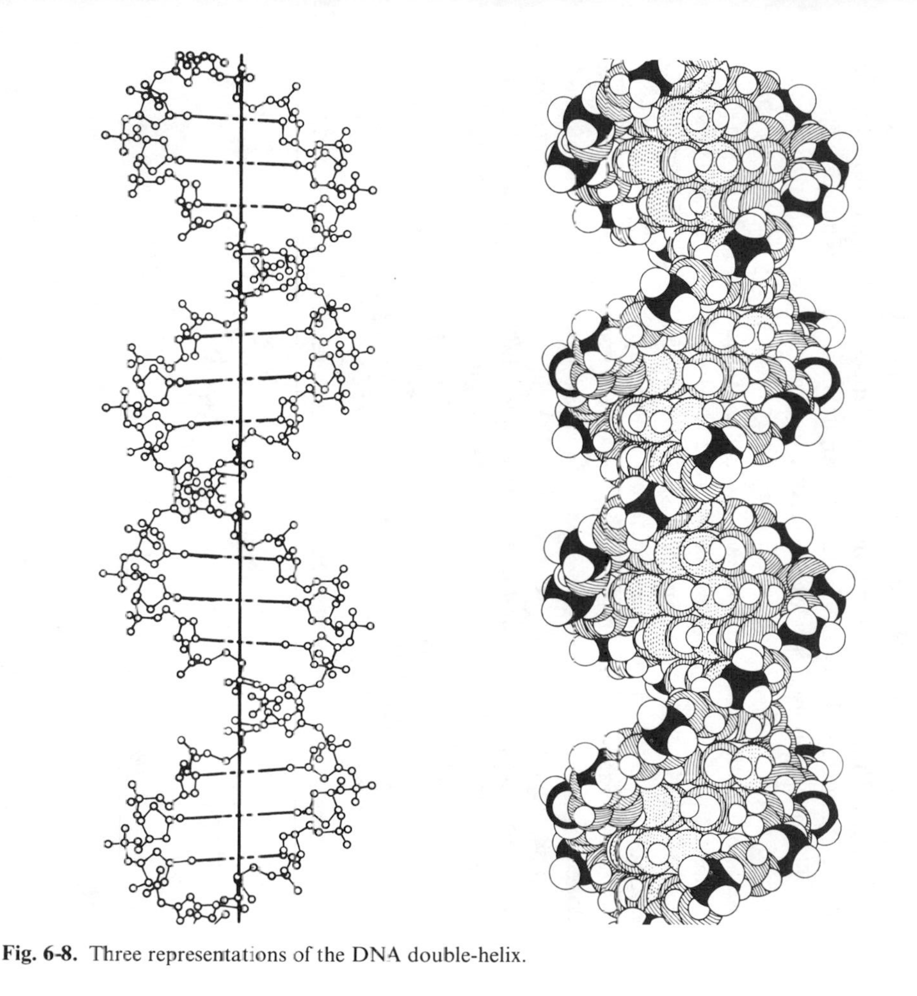

Fig. 6-8. Three representations of the DNA double-helix.

their genetic material, deoxy-5-hydroxymethyl-cytidylic acid replaces cytidylic acid. But this and other minor differences in the structure of the deoxyribonucleotides of DNA are not important in the present context.

The deoxyribonucleotides are linked together in the DNA polymer, as shown in Fig. 6-7, to form unbranched chains which may have many thousands of members. All nucleic acid molecules are formed by linking 3′- and 5′-positions of adjacent sugar residues by a phosphodiester bond. The structure of a DNA chain is completely specified when the sequence of deoxynucleotides is given. The genetic information carried by DNA resides in the nucleotide sequence; DNA is a genetic message written in a four-letter code.

The most remarkable property of DNA, however, is only recognized when we examine the structure of two-stranded DNA, the famous Watson-Crick structure. Two-stranded DNA has the form of a double-spiral, as

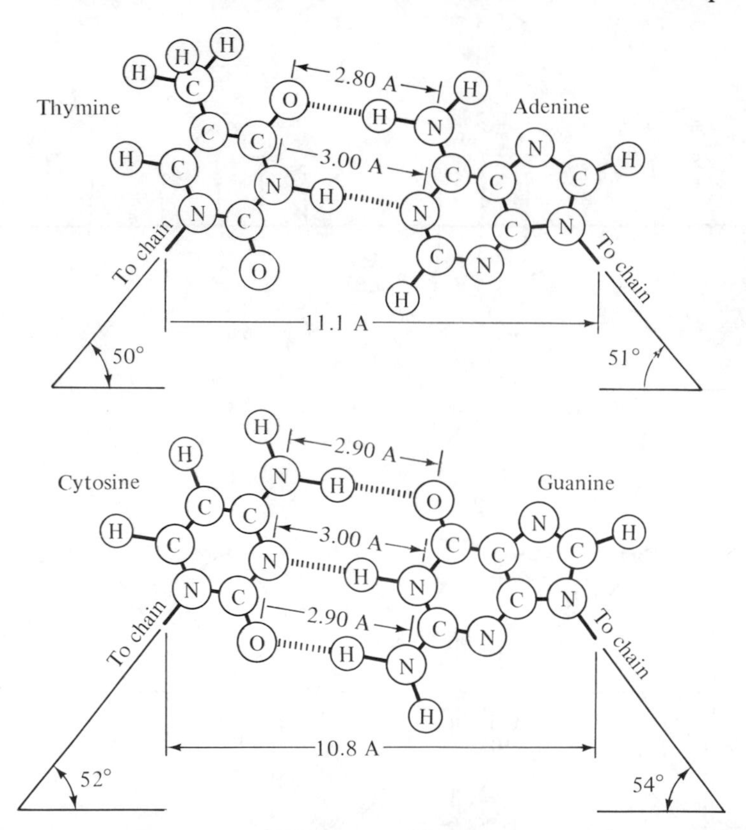

Fig. 6-9. The hydrogen-bonded base pairs in DNA. Adenine is always attached to thymine by two hydrogen bonds, whereas guanine always bonds to cytosine by three hydrogen bonds. The obligatory pairing of the smaller pyrimidine with the larger purine allows the two sugar phosphate backbones to have identical helical configurations. All the hydrogen bonds in both base pairs are strong since each hydrogen atom points directly at its acceptor atom (nitrogen or oxygen).

shown in Fig. 6-8. The strands are held together by specific hydrogen-bonding between A and T and between G and C, as shown in Fig. 6-9. The regular structure is possible only if A in one chain is bonded to T in the other (or G in one to C in the other). While hydrogen bonding between other pairs of bases is possible, it inevitably leads to nucleotide pairs which have the wrong external geometry and do not fit into the regular double-helical structure. It follows that a sequence of nucleotides on one DNA strand determines the sequence on the other strand unambiguously; for example, the sequence AGCTTC on one chain specifies the sequence GAAGCT on the other. The sequences are always given reading from the 5′ end towards the 3′ end (see Fig. 6-10).

In order to fulfill its function as a genetic material, DNA must be capable both of replication and expression. The former is achieved by the Watson-Crick mechanism. Briefly stated, the mechanism is that each strand forms

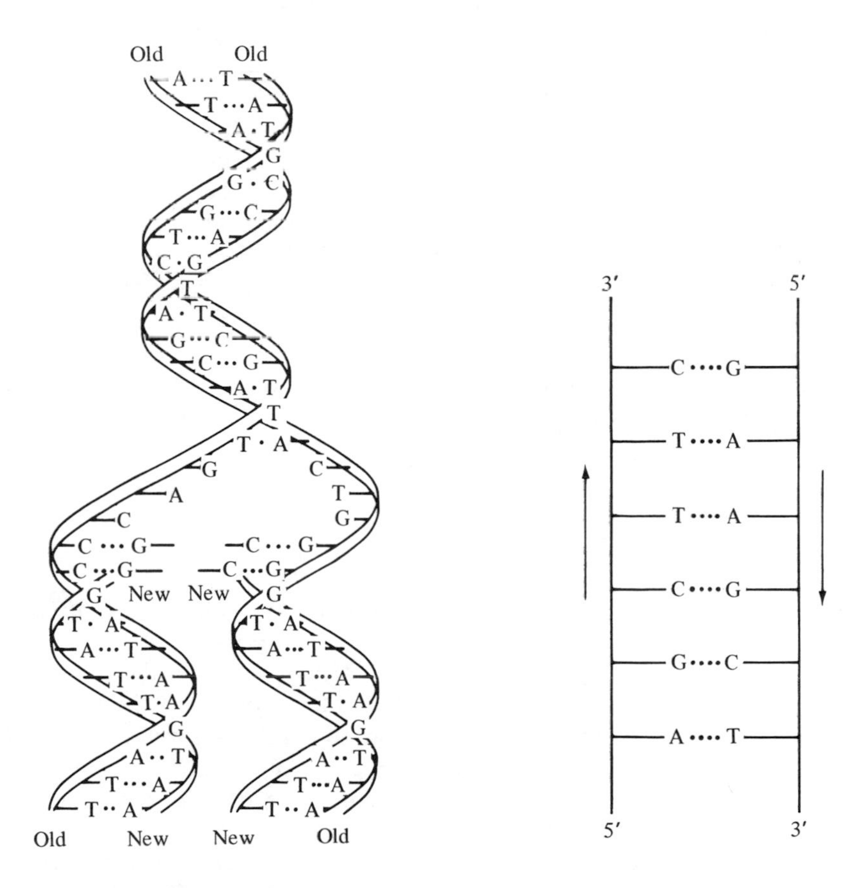

Fig. 6-10. The replication of DNA. (From Watson, 1970.)

the template for the condensation of a complementary strand. In the presence of an enzyme, deoxyribonucleotide triphosphates polymerize on a preformed DNA primer to give a new DNA strand complementary to the primer (Fig. 6-10).

It should be recognized that the feature of the nucleic acid structure which makes it ideal for a genetic material is the ability of one chain to specify exactly the sequence of its complement. Although DNA is preserved in the two-stranded form in all higher organisms, this is not essential for its replication. In certain small viruses the genetic DNA is preserved in single-stranded form, and the usual two-stranded form is synthesized only after infection. The genetic constancy of the virus is still assured, provided base-pairing is sufficiently accurate. This applies also to the replication of the genetic RNA present in a variety of single-stranded RNA viruses. Often the two-stranded form of RNA has only a transitory existence.

The expression of the genetic information encoded in the DNA is a more complex process involving a number of enzymes as well as the other principle nucleic acid, ribonucleic acid. The structure of ribonucleic acid is very closely related to that of a single DNA strand; the main difference is the replacement of deoxyribose by ribose throughout, and a second difference is the replacement of thymine by uracil.

One species of RNA, known as *messenger RNA* (m-RNA) acts as an intermediate information carrier between DNA and the protein-synthesizing machinery. It appears that, *in vivo*, DNA never directs protein synthesis itself, but rather directs the synthesis of a strand of m-RNA complementary to one of the DNA strands. In this synthesis of RNA, the same base-pairing, apart from the substitution of U for T, occurs as in DNA replication so that the RNA strand formed is complementary to the template strand of DNA in the sense of Fig. 6-10. It is this faithful copy of the DNA that is used as the final "tape" directing the synthesis of proteins. In RNA viruses, the viral RNA (or its complementary strand) serves both as genetic material and as messenger RNA.

PROTEIN SYNTHESIS

The final transcription of the RNA message takes place on a complex particle called a *ribosome*. The ribosome is made up of a special ribosomal RNA (60%) and a group of proteins (40%). The function of the ribosomal RNA is unknown, as are many details of protein synthesis. However, the distinction between messenger RNA and ribosomal RNA is well-established. Messenger RNA carries information for protein sequences; ribosomal RNA is part of the translation apparatus. The following description of protein synthesis is much oversimplified.

Prior to their assembly into proteins, the amino acids are activated to form enzyme-bound aminoacyladenylates. There is usually one enzyme

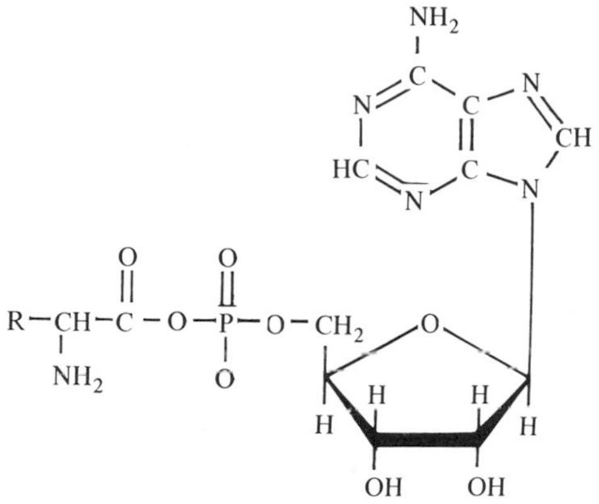

Fig. 6-11. The structure of aminoacyl adenylates. They are synthesized enzymatically from ATP and amino acids. The same enzymes transfer the amino acid to the appropriate t-RNA.

for each amino acid, the interaction of an amino acid with its activating enzyme being highly specific. The amino acids after activation are next transferred to the terminal nucleotide of another special RNA, a low molecular weight RNA known as *transfer RNA* (t-RNA). The t-RNAs do not carry genetic information. For each amino acid there is one or a small number of t-RNAs, and the transfer from the activating enzyme to the t-RNA is again highly specific. Once the amino acid is attached to the t-RNA, it is incorporated into protein without further attention to the nature of the amino acid. There is one exception to this rule. Valine may become attached to the isoleucine t-RNA but it is then hydrolyzed off before it can be incorporated into protein.

The stage is now set for the final synthesis of a protein: m-RNA, and the amino acids, prepared for condensation by attachment to t-RNA, come together in the ribosome. Each sequence of three nucleotides on the m-RNA (codon) corresponds either to just one "recognition" site on a particular t-RNA (anticodon), or to a stop sign. The ribosome is so constructed that it slips along the messenger tape in jumps of three bases at a time; at each step the polypeptide chain is extended by adding the appropriate amino acid. The manner in which the growing polypeptide chain is handed on from one RNA molecule to another is illustrated in Fig. 6-12.

The genetic code is given in Table 6-1. It should be noted that a few of the triplets do not stand for amino acids, but rather for stop signs—that is, signs that the amino acid sequence of a protein is complete and that the protein should be detached from the ribosome. Similarly, there is a special codon for the amino acid which begins a protein.

Before completing this very cursory account of molecular genetics, we must return to the RNA viruses. These viruses use the same strand of RNA

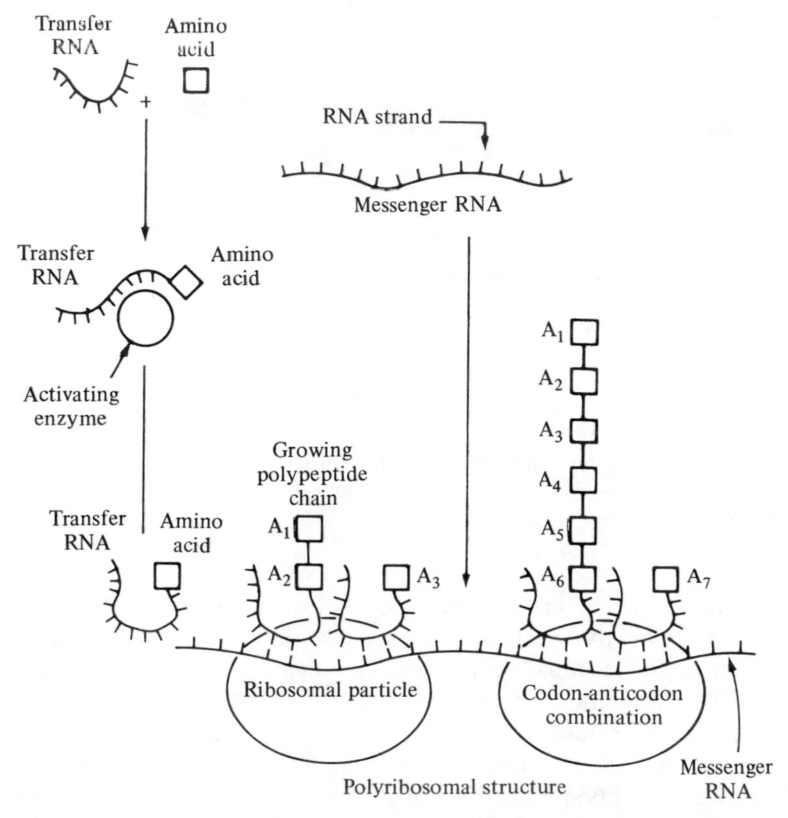

Fig. 6-12. A schematic outline of protein synthesis. The amino acid is attached to the transfer RNA after activation by ATP giving an amino acyl-t-RNA molecule. The amino acyl-t-RNAs enter the ribosome and the anticodon pairs with the codon of the messenger RNA. The growing peptide chain is transferred to the amino group at the entering amino acid. Note that two polypeptide chains are growing on a single messenger, RNA.

both as a template for RNA synthesis and as a messenger for protein synthesis. Thus, they represent organisms with a simplified genetic structure. Of course their existence does not establish that RNA replication is more primitive than DNA replication, but it does suggest strongly that a self-sufficient genetic system based on RNA and protein is possible. It is also possible that DNA alone could have done the job, provided a terminal 2′ OH group is not essential for t-RNA function.

The sequence of a number of t-RNAs have now been determined. Since they are peculiarly important in any consideration of the evolution of the code, we shall discuss them in a little more detail. The sequence of a typical t-RNA, the alanyl-t-RNA of yeast, is given in Fig. 6-13. One of the features of this sequence is the presence of a number of modified bases, for example, the methylated bases, dihydrouracil, and pseudouridine. The function of

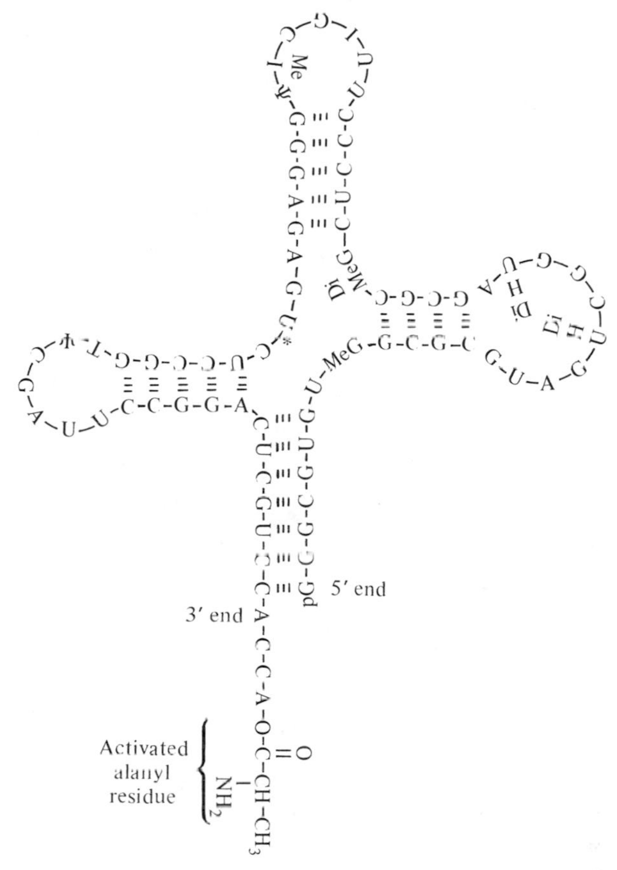

Fig. 6-13. The structure of alanine t-RNA. The anticodon is IGC. The amino acid is attached to the terminal A. DiHU, dihydrouridine; DiMeG, 2-dimethyl guanine; MeI, 1-methylinosine; ψ, pseudouridine; U*, a mixture of U and DiHU. (From Holley *et al.*, *Science* **147**, 1462, 1965.)

these minor components of t-RNA is not understood, but is believed to be concerned with maintaining a complicated three-dimensional structure.

A little more must now be said about the anticodons in the t-RNAs. If the standard rules applied for pairing of bases in the anticodon with the corresponding triplet in the messenger, then we would need as many t-RNAs as there are codons for an amino acid (1 to 6). However, the structure of the genetic code is such that variations in the third base of a codon are often without effect on the amino acid coded for (Table 6-1). It would obviously be economical to use a single base in the third position of the anticodon to recognize an appropriate set of bases in the codon, since one t-RNA would then be needed instead of two or more. This is just what is found, for while the anticodon-codon recognition is by standard pairing in the first two positions, the rules are more complicated for the third position.

Table 6-1. The Genetic Code[a]

SECOND LETTER

FIRST LETTER	U	C	A	G	THIRD LETTER
U	UUU } Phe	UCU } Ser	UAU } Tyr	UGU } Cys	U
	UUC	UCC	UAC	UGC	C
	UUA } Leu	UCA	UAA OCHRE	UGA UMBER	A
	UUG	UCG	UAG AMBER	UGG Tryp	G
C	CUU } Leu	CCU } Pro	CAU } His	CGU } Arg	U
	CUC	CCC	CAC	CGC	C
	CUA	CCA	CAA } GluN	CGA	A
	CUG	CCG	CAG	CGG	G
A	AUU } Ileu	ACU } Thr	AAU } AspN	AGU } Ser	U
	AUC	ACC	AAC	AGC	C
	AUA	ACA	AAA } Lys	AGA } Arg	A
	AUG Met	ACG	AAG	AGG	G
G	GUU } Val	GCU } Ala	GAU } Asp	GGU } Gly	U
	GUC	GCC	GAC	GGC	C
	GUA	GCA	GAA } Glu	GGA	A
	GUG	GCG	GAG	GGG	G

[a] UAA (ochre), UAG (amber), and UGA (umber) are chain terminating codons. AUG is used as a chain initiating codon standing for formylmethionine in E. coli. In the middle of a protein chain it stands for methionine.

Table 6-2. The "Wobble" Hypothesis. This has been proposed to provide rules for the pairing of codon and anticodon at the third position of the codon. There is evidence, for example, that the anticodon base I, which stands for inosine, may pair with as many as three different bases: U, C, and A. Inosine closely resembles the base guanine (G) and so would ordinarily be expected to pair with cytosine (C).

Anticodon	Codon
U	A G
C	G
A	U
G	U C
I	U C A

The most interesting feature of pairing in this third position is the frequent occurrence of inosinic acid, which is able to pair with U, C, and A. Further details of the allowed pairings are collected in Table 6-2. Clearly, it is a matter of great importance to discover which of these details of contemporary protein synthesis are primitive and which recent. For example, are t-RNAs primitive? Was inosine present from the beginning or was it incorporated into t-RNA only after the evolution of the ribosome?

METABOLISM

Although the genetic apparatus of different organisms is always based on a self-replicating nucleic acid and universal RNA-protein translation mechanism, the supporting metabolic processes vary from one organism to another. Some contemporary organisms obtain their energy by oxidizing carbohydrates, others by anaerobic utilization of a variety of inorganic and organic molecules, and still others obtain energy directly from sunlight by photophosphorylation. A closer study, however, reveals an underlying unity in these processes.

Almost all of the energy transactions of any cell are carried out using "high-energy" phosphate compounds. These are compounds such as adenosine triphosphate (ATP) and other nucleoside triphosphates, which release about 8 kcal of free energy when they hydrolyze to inorganic phosphate. ATP is certainly the most important of these compounds, and our later discussion will be restricted to it.

The proteins and nucleic acids are built from their component amino acids, sugars, bases, and phosphates by dehydration. This is possibly only

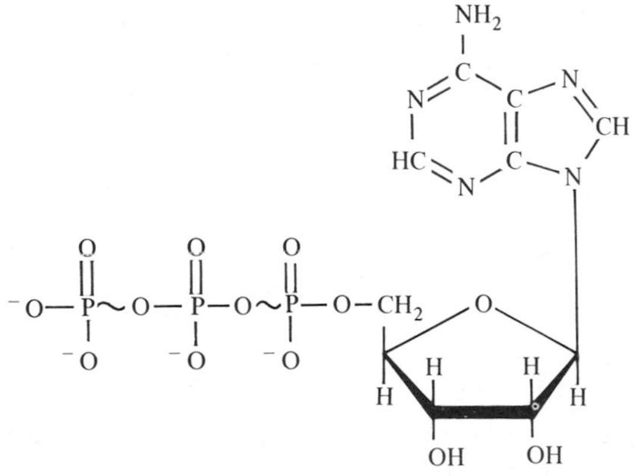

Fig. 6-14. The structure of ATP. The wavy lines represent bonds which release about 8 kcal of free energy when they hydrolyze.

at the cost of chemical free energy. The equilibrium state in aqueous solutions of peptides and polynucleotides at physiological pH is one of almost complete hydrolysis. The direct source of this free energy is a high-energy phosphate compound, almost always ATP. Thus, while organisms are very different in the preliminary steps by which they make ATP, once they have made it they are all very similar in the way they use it to synthesize proteins and nucleic acids. One should add, for completeness, that contemporary organisms use ATP for many purposes other than those discussed above (for some steps in the biosynthesis of amino acids and nucleotide bases, in muscle contraction, and in moving molecules against a concentration gradient, for example).

The participation of ATP in nucleic synthesis involves the synthesis of the various triphosphates from the appropriate deoxynucleotides or nucleotides by phosphate transfer, followed by the condensation of the triphosphates to give DNA or RNA and inorganic pyrophosphate. The net result is the consumption of ATP and the production of inorganic pyrophosphate; the hydrolysis of ATP is coupled to the dehydration of nucleotides to polynucleotides. In a precisely parallel fashion, the formation of proteins from amino acids is a dehydration reaction coupled to the hydrolysis of ATP via the formation of aminoacyladenylates (Fig. 6-11).

$$\text{ATP} + \text{amino acid} \rightleftarrows \text{pyrophosphate} + \text{aminoacyladenylate}$$

$$n \text{ aminoacyladenylates} \longrightarrow (\text{amino acid})_n + n \text{ adenylic acid (AMP)}$$

In addition, GTP is hydrolyzed in the ribosome during peptide synthesis.

It should be clear that one of the crucial questions concerning the origins of life is the nature of the primitive dehydrating agent. Was it ATP? If so, where did it come from? If not, what other compound or compounds could have substituted for ATP? Alternatively, could the necessary dehydrations have occurred as thermal or photochemical processes? (See Chapter 11.)

A great deal of discussion has been devoted to metabolic pathways in the origin of life. It should be noted that while the genetic mechanism is universal, biosynthetic pathways are occasionally different in different organisms. The processes used in energy production are even more variable. This makes it more difficult to argue backwards from present organisms to the metabolism of the earliest organism (see Chapter 14). Hopefully, one can argue backwards for the genetic apparatus, although even this approach has its dangers.

MEMBRANES

All cells are surrounded by a membrane composed of phospholipids and proteins. These membranes prevent useful molecules from diffusing out of the cell or harmful molecules from diffusing in. Cell membranes also contain enzyme-like proteins which can transport selected molecules from the outside of the cell to the inside against a concentration gradient. Naturally, this requires an expenditure of energy, the source of which is the hydrolysis of ATP.

Most cells have a potential difference ranging from 10 to 100 millivolts between the inside and the outside, with the outside positive. The changes of this potential in nerve cells results in the conduction of a signal. Furthermore a substantial part of the metabolism of a cell takes place on membranes.

These considerations have led some workers to maintain that membranes, being so important in present day organisms, were essential for the first living organism. The most important role postulated for primitive membranes in the origin of life is that of keeping essential molecules from diffusing away. It seems unlikely that primitive membranes would have been sufficiently developed to carry out active transport.

There are a number of alternatives to the use of a membrane for keeping organic molecules together. It is possible that the concentration of organic compounds wherever life evolved was so great that no membranes were needed, or alternatively, that the molecules were kept together in a nonaqueous phase such as a drop of oil or a coacervate (see Chapters 10 and 12). Another possibility is that the important molecules were adsorbed on a clay mineral, on hydroxylapatite, or on some other solid particle. Absorption at the air-water interface at the surface of the ocean is another possibility.

Since these alternatives are available, it is not certain that membranes were used very early. On the other hand, artificial lipid membranes are produced so easily from solutions of lipids that if lipids were synthesized on the primitive earth, they would very likely have formed simple membranes. Some of these might have served to keep assemblies of prebiotic molecules together. It remains to be determined whether such membranes did play a significant role in the organization of the first living cell.

MUTATION AND NATURAL SELECTION

In our discussion of the contemporary genetic system we have described a nucleic acid blueprint which replicates accurately and guides the other synthetic activities of the cell. Such a description gives no account of the evolution of biological systems. In fact, there is one widely accepted mechanism which accounts for all, or nearly all, of the genetic changes which are observed in simple organisms.

The replication of nucleic acids, while extremely accurate by all normal standards, is not perfect. Under normal circumstances there is a probability of about 10^{-8} for each base that during replication of DNA a mistake occurs, and instead of the complementary base one of the other three is substituted. For a typical bacterial population this would give one mistake in every 10–100 daughter cells. Other, more complicated changes—for example, deletions of long strings of nucleotides—can also occur. It is often possible to boost the mutation rate by many orders of magnitude by subjecting the cell to ultraviolet or high-energy radiation, or by treating it with certain reagents known as *mutagens*.

The most common consequence of a mutational change is the production of protein containing an altered amino acid, although many other kinds of changes occur less frequently. The new protein may be inactive, have a somewhat modified activity, or, on rare occasions, acquire a new activity. Since the relation between the amino acid change and the "symptoms" that such change produces may be very indirect in a complex organism, it is best, for our purposes, to distinguish between three very broad classes of effects that genetic changes may have on the characteristic appearance and behavior of a complex organism.

1. The organism's ability to deal with its environment is impaired.
2. The organism's ability to deal with its environment is unchanged.
3. The organism's ability to deal with its environment is improved.

In case (1), the defect (under natural conditions) is soon eliminated from the population through the competition of unchanged and therefore more efficient organisms. In case (2), the result is often the formation of a stable, mixed population. In case (3), there is a good chance that the modified organisms outgrow all competitors and in a short time establish themselves as the dominant variety.

All the organisms with which we are familiar have a long evolutionary history and are therefore adapted to their environment. Under these circumstances almost all detectable mutations are bad mutations (case 1): Most of the useful ones have already been incorporated in the DNA (neutral mutations are unlikely to be detected). It is only when the environment changes suddenly that we are likely to see a rapid evolutionary adaptation. Many examples are known in the realm of microbiology. The development of resistance to antibiotics is a well-known example of the selection of mutants in an altered environment.

We believe that the concept of evolution by mutation and natural selection is the key to an understanding of the origin of life. The greatest contribution of molecular biology to the understanding of the origin of life is the realization that the emergence of biological order was neither more nor less than the production of imperfectly replicating polymers on which natural selection could act.

References

This chapter summarizes the material to be found in most molecular biology or modern biology textbooks. A particularly widely used book is J. D. Watson, *Molecular Biology of the Gene*, 2nd ed. (Benjamin, New York, 1970).

Chapter Seven

Prebiotic Synthesis of Amino Acids, Urea, Fatty Acids, Porphyrins and Vitamins

THE SYNTHESIS OF AMINO ACIDS

Amino acids were the first biologically interesting organic compounds to be identified as products formed under simulated primitive earth conditions. This was partly because amino acids were more easily identified at the time than purines and pyrimidines, and partly because they were formed in a single continuous set of operations under reasonably plausible prebiotic conditions.

The apparatus used in the original experiments is shown in Fig. 7-1. Water was first added to the small flask and the air was pumped out. Then a mixture of ammonia, methane and hydrogen was added. The water in the 500 ml flask was boiled to bring the vapor into the region of the spark and to circulate the gases in the 5 liter flask. An electric spark was generated across the spark gap by a Tesla coil of the type used in testing for leaks in vacuum systems.

The products of the discharge were condensed by the condenser and washed through the U-tube into the small flask. The nonvolatile products remained there, but the volatile products recirculated past the spark. This apparatus was claimed to be a crude model of the primitive earth. The reduced forms of carbon, nitrogen, and oxygen in the gas phase represented the atmosphere, the liquid phase represented the ocean.

The spark was operated continuously for a week, after which time the gases were pumped out and the accumulated products in the small flask analyzed by chromatography. Sufficient quantities of most of the compounds identified were obtained to prepare derivatives and determine their melting points.

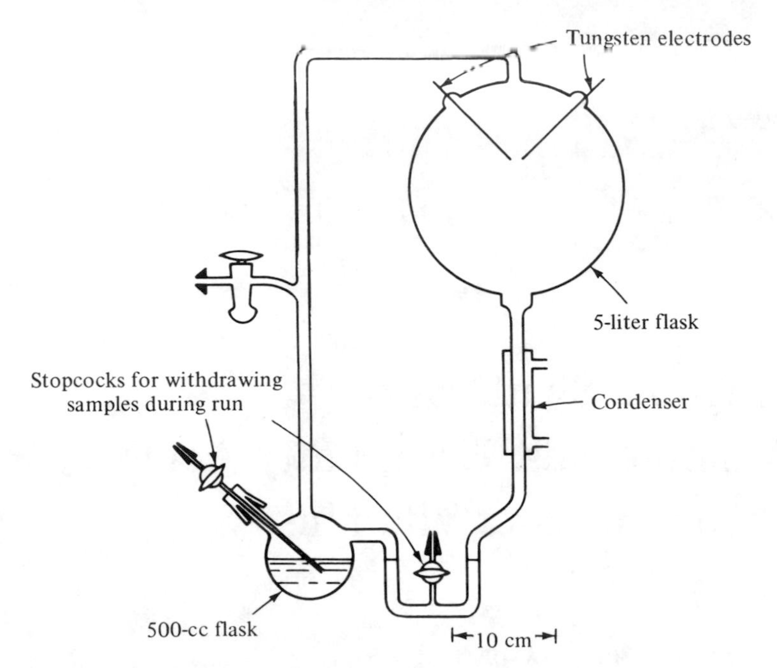

Fig. 7-1. Apparatus for the electric discharge synthesis of amino acids.

The results of this analysis are given in Table 7-1. The compounds identified account for 15% of the carbon added to the apparatus. A substantial quantity of polymer or tar was found, perhaps a cyanide, aldehyde, or mixed cyanide-aldehyde polymer. In addition, some of the methane remained undecomposed in the gas phase along with CO and N_2 formed in the spark.

The results of this experiment were unexpected in two respects. A small number of relatively simple compounds accounted for a large proportion of the products. A very complicated mixture containing small amounts of a great variety of compounds would have been anticipated. Furthermore, the major products were not themselves a random selection of organic compounds but included a surprising number of substances that occur in living organisms.

The mechanism of synthesis of amino acids was studied in some detail. The concentrations of NH_3, HCN, and aldehydes in the U-tube and the concentrations of amino acids in the 500 ml flask were determined at intervals during a run that lasted for a week. The results are shown in Fig. 7-2. These data show that large amounts of cyanide and aldehydes were formed during the first 125 hours of sparking and that the rate of their synthesis then fell off. The amino acids were formed at a more or less constant rate throughout the run.

Table 7-1. Yields from sparking a mixture of CH_4, NH_3, H_2O, and H_2[a]

Compound	Yield (μM)	Yield (%)
Glycine	630	2.1
Glycolic acid	560	1.9
Sarcosine	50	0.25
Alanine	340	1.7
Lactic acid	310	1.6
N-Methylalanine	10	0.07
α-Amino-n-butyric acid	50	0.34
α-Aminoisobutyric acid	1	0.007
α-Hydroxybutyric acid	50	0.34
β-Alanine	150	0.76
Succinic acid	40	0.27
Aspartic acid	4	0.024
Glutamic acid	6	0.051
Iminodiacetic acid	55	0.37
Iminoaceticpropionic acid	15	0.13
Formic acid	2,330	4.0
Acetic acid	150	0.51
Propionic acid	130	0.66
Urea	20	0.034
N-Methyl urea	15	0.051

[a] 59 mmoles (710 mg) of carbon was added as CH_4. The per-cent yields are based on the carbon.

Hydrogen cyanide, aldehydes, and ammonia are known to react as shown in the following equations.

$$\mathrm{RCHO + HCN + NH_3 \rightleftarrows \underset{\displaystyle NH_2}{\underset{|}{RCH}}{-}C{\equiv}N + H_2O}$$

$$\mathrm{\underset{\displaystyle NH_2}{\underset{|}{RCH}}{-}C{\equiv}N + 2H_2O \longrightarrow \underset{\displaystyle NH_2}{\underset{|}{RCH}}{-}COOH + NH_3}$$

It was shown that most, if not all, of the amino acids produced in the electric discharge apparatus could have been formed in this way. We shall discuss the details of these reactions in a later section. Here we note that a mechanism of this type, involving condensation of simple active intermediates in aqueous solution, readily accounts for the limited variety of products obtained. If more complex compounds were formed directly in the discharge, far more different substances would be obtained by the recombination of radicals and ions formed in the spark.

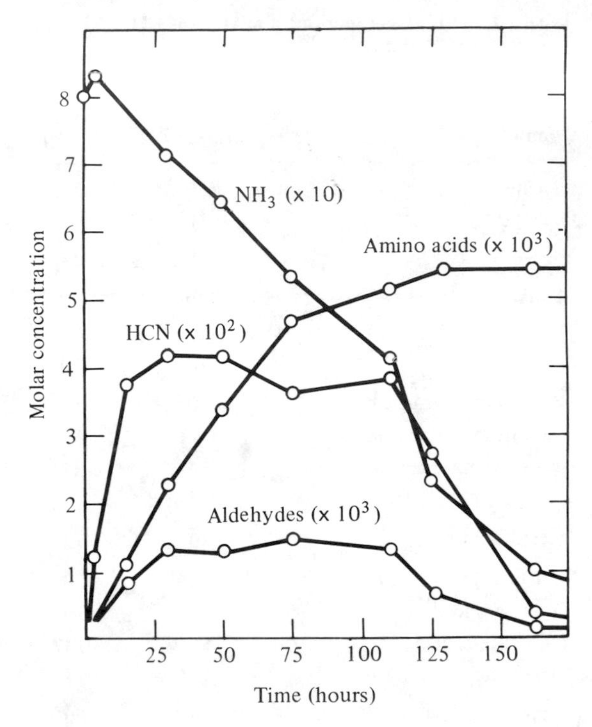

Fig. 7-2. The concentrations of ammonia, hydrogen cyanide, and aldehyde in the U-tube, and the concentration of amino acids in the 500 ml flask while sparking a mixture of methane, ammonia, water, and hydrogen in the apparatus shown in Fig. 7-1.

A variety of related experiments have now been performed using different energy sources, such as silent discharges or ultraviolet light. The results have usually been qualitatively similar but the quantitative yields vary considerably from one set of conditions to another. Amino acids are also produced if the original CH_4-NH_3-N_2 mixture is replaced by other reducing mixtures—for example, CO-N_2-H_2.

The electric discharge amino acid synthesis has been repeated recently using a mixture of CH_4, N_2, and H_2O, with only traces of ammonia. The analysis of products was performed using modern and more sensitive techniques than the early experiments. The results of a typical run are shown in Table 7-2. It should be noted that glycine, alanine, and α-aminobutyric acid are obtained in considerably greater yield than valine, leucine, and isoleucine. Some amino acids which do not occur in proteins (e.g., norvaline and norleucine) are obtained in comparable yields. The nonprotein amino acids (e.g., N-ethyl glycine, α-aminoisobutyric acid, and isovaline) are of interest because some of them have been found in the Murchison meteorite (see Chapter 14).

The action of ultraviolet light with wavelengths of 1,296 and 1,470 Å on a mixture of CH_4 (and/or C_2H_6), NH_3, and H_2O gives only very low yields

Table 7-2. Yields and mole ratios of amino acids from sparking 336 M of CH_4[a]

Compound	Yield (μM)	Compound	Yield (μM)
Glycine	440	α,γ-Diaminobutyric acid	33
Alanine	790	α-Hydroxy-γ-aminobutyric acid	74
α-Amino-n-butyric acid	270	Sarcosine	55
α-Aminoisobutyric acid	~30	N-Ethylglycine	30
Valine	19.5	N-Propylglycine	~2
Norvaline	61	N-Isopropylglycine	~2
Isovaline	~5	N-Methylalanine	~15
Leucine	11.3	N-Ethylalanine	<0.2
Isoleucine	4.8	β-Alanine	18.8
Alloisoleucine	5.1	β-Amino-n-butyric acid	~0.3
Norleucine	6.0	β-Amino-isobutyric acid	~0.3
tert-Leucine	<0.02	γ-Aminobutyric acid	2.4
Proline	1.5	N-Methyl-β-alanine	~5
Aspartic acid	34	N-Ethyl-β-alanine	~2
Glutamic acid	7.7	Pipecolic acid	~0.05
Serine	5.0	α,β-Diaminopropionic	6.4
Threonine	~0.8	Isoserine	5.5
Allothreonine	~0.8		

[a] The yields of glycine and alanine, based on the carbon, are 0.26% and 0.71%, respectively. The total yield of amino acids in the table is 1.90%.

of amino acids. The yields can be considerably improved by adding H_2S to the above mixture of gases. Light with a wavelength as long as 2,537 Å is effective in this synthesis. Since much more energy from the sun is available at wavelengths below 2,600 Å than is available below 1,500 Å, this type of ultraviolet light synthesis may have made a major contribution to prebiotic amino acid synthesis. Cysteine was reported as a product in this photochemical reaction, in addition to glycine, alanine, aspartic acid, glutamic acid, and serine.

A second potentially prebiotic synthesis of amino acids was demonstrated by Oró, who showed that if solutions of ammonium cyanide are refluxed and the product subjected to acid hydrolysis, glycine is obtained in several percent yield. In addition, smaller quantities of aspartic acid, serine, and alanine are formed along with traces of other amino acids. Clearly, this mechanism could contribute to the yield of amino acids in the discharge experiments and in other syntheses which yield hydrogen cyanide in large amounts.

Rather than discuss the details of the many amino acid syntheses that have been reported, we shall describe what is known about the mechanisms of the two major syntheses.

THE STRECKER SYNTHESIS

The Strecker synthesis is the oldest organic synthesis of amino acids. The first stage is the addition of ammonia to an aldehyde to give an imine.

$$R{-}\overset{\displaystyle O}{\overset{\|}{C}}{-}H + NH_3 \rightleftarrows R{-}\underset{\displaystyle NH_2}{\underset{|}{\overset{\displaystyle OH}{\overset{|}{C}}}}{-}H \rightleftarrows R{-}\underset{\displaystyle NH}{\underset{\|}{C}}H + H_2O \qquad (1)$$

The cyanide then adds to the imine to give an aminonitrile.

$$R{-}\underset{\displaystyle NH}{\underset{\|}{C}}{-}H + HCN \rightleftarrows R{-}\underset{\displaystyle NH_2}{\underset{|}{C}}H{-}C{\equiv}N \qquad (2)$$

These steps are reversible. The reaction is completed by the irreversible hydrolysis of the nitrile.

$$R{-}\underset{\displaystyle NH_2}{\underset{|}{C}}H{-}C{\equiv}N \xrightarrow{H_2O} R{-}\underset{\displaystyle NH_2}{\underset{|}{C}}H{-}\overset{\displaystyle O}{\overset{\|}{C}}{-}NH_2 \xrightarrow{H_2O} R{-}\underset{\displaystyle NH_2}{\underset{|}{C}}H{-}COOH + NH_3 \qquad (3)$$

A parallel series of reactions leads to the formation of hydroxy-acids.

$$R{-}\underset{\displaystyle O}{\underset{\|}{C}}{-}H + HCN \rightleftarrows R{-}\underset{\displaystyle OH}{\underset{|}{C}}H{-}C{\equiv}N \qquad (2')$$

$$R{-}\underset{\displaystyle OH}{\underset{|}{C}}H{-}C{\equiv}N \xrightarrow{H_2O} R{-}\underset{\displaystyle OH}{\underset{|}{C}}H{-}\overset{\displaystyle O}{\overset{\|}{C}}{-}NH_2 \xrightarrow{H_2O} R{-}\underset{\displaystyle OH}{\underset{|}{C}}H{-}COOH + NH_3 \qquad (3')$$

Equilibrium constants for the reversible formation of aminonitriles and hydroxynitriles may be written as follows:

$$RCHO + NH_3 + HCN \rightleftarrows R\underset{\displaystyle NH_2}{\underset{|}{C}}HCN + H_2O$$

$$K = \frac{[RCH(NH_2)CN]}{[RCHO][HCN][NH_3]} \qquad (4)$$

$$RCHO + HCN \rightleftarrows R\underset{\displaystyle OH}{\underset{|}{C}}HCN \qquad H = \frac{[RCH(OH)CN]}{[RCHO][HCN]} \qquad (5)$$

The equilibrium between the aminonitrile and hydroxynitrile

$$RCH(OH)CN + NH_3 \rightleftarrows RCH(NH_2)CN + H_2O$$

is then governed by the relation

$$\frac{[RCH(NH_2)CN]}{[RCH(OH)CN][NH_3]} = \frac{K}{H} \tag{6}$$

Expression 6 shows that if the ammonia concentration is high, the aminonitrile is the major product; otherwise, the hydroxynitrile is formed in large quantities. In typical experiments, the aminonitrile and hydroxynitrile are obtained in equal amounts when the concentration of free ammonia (not the ammonium ion) is about 0.01 M at 25°C.

Values of H for typical aldehydes range from 10^2 to 10^6. This means that equal amounts of aldehyde and hydroxynitrile will be present at cyanide concentrations varying from 10^{-2} to 10^{-4} M, depending on the aldehyde. In the presence of excess ammonia similar concentrations of cyanide are required to produce the aminonitrile.

The hydrolysis reactions 3 and 3′ are irreversible. These rates have not been compared experimentally, but theory suggests that the aminonitrile should hydrolyze faster than the hydroxynitrile, thus favoring the synthesis of amino acids.

In synthetic organic chemistry the Strecker synthesis is usually completed by hydrolysis of the aminonitrile in strong acid (although alkali is sometimes used). The hydrolysis in the primitive ocean (pH = 8.0) would have been slow, especially if the temperature was low. Nonetheless, this is not a serious problem since the reaction would certainly be complete in 10,000 years, a small part of the time available for prebiotic synthesis.

The Strecker synthesis requires, in addition to cyanide, which is obtained quite readily, appropriate aldehydes—e.g., formaldehyde, acetaldehyde, and glycolaldehyde as precursors of glycine, alanine, and serine, respectively. The simple aliphatic aldehydes are obtained by the action of electric discharges or ultraviolet light on methane-water mixtures. The mechanisms of these syntheses are not understood. Ultraviolet light seems more efficient for the synthesis of simple aliphatic aldehydes, but discharges produce more hydrogen cyanide.

The aldehyde precursors of the more complicated amino acids can be synthesized by aldol condensation reactions in solution, by reactions in the gas phase, and by hydration of activated acetylenes ($R{-}C{\equiv}CH + H_2O \longrightarrow RCH_2{-}CHO$). Glycolaldehyde, for example, can be synthesized by sparking mixtures of methane and water and also by the condensation of two molecules of formaldehyde (see section on sugars in Chapter 8).

$$HCHO + HCHO \longrightarrow \underset{\displaystyle OH}{\underset{|}{C}H_2}{-}CHO$$

Acrolein can be obtained by the reaction of formaldehyde with acetaldehyde:

$$HCHO + CH_3CHO \rightarrow \underset{\displaystyle OH}{\underset{|}{C}}H_2{-}CH_2CHO \rightarrow CH_2 = CH{-}CHO + H_2O$$

Acrolein can also be obtained directly by sparking a mixture of methane and water in fair yield (0.04% based on the carbon). Acrolein reacts in dilute aqueous solution to give at least five amino acids, two of which occur in proteins.

$$CH_2{=}CH{-}CHO \xrightarrow{CH_3SH} CH_3S{-}CH_2{-}CH_2{-}CHO \xrightarrow[\text{hydrolysis}]{NH_3\ \ HCN} \text{Methionine}$$

$$CH_2{=}CH{-}CHO \xrightarrow{H_2S} HS{-}CH_2{-}CH_2{-}CHO \longrightarrow \text{Homocysteine}$$

$$CH_2{=}CH{-}CHO \xrightarrow{HCN} NC{-}CH_2{-}CH_2{-}CHO \longrightarrow \text{Glutamic acid}$$

$$CH_2{=}CH{-}CHO \xrightarrow{H_2O} HO{-}CH_2{-}CH_2{-}CHO \longrightarrow \text{Homoserine}$$

$$CH_2{=}CH{-}CHO \xrightarrow{NH_3} H_2N{-}CH_2{-}CH_2{-}CHO \longrightarrow \alpha,\gamma\text{-Diamino-butyric acid}$$

The addition of CH_3SH and H_2S to give methionine and homocysteine is greatly favored over the addition of the other nucleophiles. Methionine can be produced in small yield (2×10^{-4}% based on the carbon) by sparking a mixture of CH_4, N_2, NH_3, and H_2S, and can be produced in substantially greater yield (6×10^{-3}% based on the carbon; 0.6% based on the sulfur) by sparking a mixture of CH_4, N_2, NH_3, and CH_3SH.

A variation of the Strecker reaction has been reported by Ponnamperuma. An arc electric discharge acting on a mixture of methane and ammonia produces the nitriles of glycine, alanine, and sarcosine ($CH_3HN{-}CH_2{-}CN$) directly. On solution in water these nitriles would equilibrate with aldehydes, ammonia cyanide, and hydroxynitriles (reactions 1 and 2). This mechanism is therefore roughly equivalent to the Strecker synthesis described above.

Synthesis of aspartic acid, phenylalanine, and tyrosine have been reported in which an activated acetylene replaces an aldehyde as the starting material in the Strecker synthesis. Cyanoacetylene is a major product (as much as 4%) of the action of an electric discharge on a N_2-CH_4 mixture. It reacts with ammonium cyanide to give the nitrile of aspartic acid which hydrolyzes to aspartic acid in quite good yield. Asparagine (and isoasparagine) are partial hydrolysis products of aspartic nitrile.

$$N{\equiv}C{-}CH_2{-}\underset{\displaystyle NH_2}{\underset{|}{C}}H{-}C{\equiv}N \xrightarrow{\text{hydrolysis}} H_2N{-}\overset{\displaystyle O}{\overset{\|}{C}}{-}CH_2\underset{\displaystyle NH_2}{\underset{|}{C}}HCOOH$$

$$\rightarrow HOOCCH_2\underset{\displaystyle NH_2}{\underset{|}{C}}HCOOH$$

The formation of the amino dinitrile could occur in several almost equivalent ways.

$$\mathrm{N{\equiv}C{-}C{\equiv}CH \xrightarrow{NH_3} N{\equiv}C{-}CH{=}CH(NH_2) \longrightarrow N{\equiv}C{-}CH_2{-}CH{=}NH \xrightarrow{HCN} N{\equiv}C{-}CH_2{-}CH(NH_2){-}C{\equiv}N}$$

$$\mathrm{N{\equiv}C{-}C{\equiv}CH \xrightarrow{HCN} N{\equiv}C{-}CH{=}CH{-}C{\equiv}N \xrightarrow{NH_3} N{\equiv}C{-}CH_2{-}CH(NH_2){-}C{\equiv}N}$$

$$\mathrm{N{\equiv}C{-}C{\equiv}CH \xrightarrow{H_2O} N{\equiv}C{-}CH_2{-}CH{=}O \xrightarrow[HCN]{NH_3} N{\equiv}C{-}CH_2{-}CH(NH_2){-}C{\equiv}N}$$

The pathway on the primitive earth would have depended on the NH_3 and HCN concentrations. In any case, cyanoacetylene can be considered as equivalent to a source of cyanoacetaldehyde.

Phenylalanine can be synthesized from phenylacetylene. The latter compound can be made under primitive earth conditions by the action of electric discharges on methane, ethane, ethylene or acetylene, or by the pyrolysis of any of these hydrocarbons. Pyrolysis in general is more efficient than the action of electric discharges; the pyrolysis of ethane with a hot wire at 1,200°C gives a 5% yield of phenylacetylene. The action of ultraviolet light on acetylene also gives phenylacetylene.

The hydration of phenylacetylene to phenylacetaldehyde by OH^- or NH_3 at plausible concentrations is a slow reaction. However, hydrogen sulfide is an excellent catalyst. A Strecker synthesis using phenylacetaldehyde gives phenylalanine.

$$\mathrm{C_6H_5{-}C{\equiv}CH \xrightarrow{HS^-} C_6H_5{-}CH{=}CH{-}SH \longrightarrow C_6H_5{-}CH_2{-}C({=}S){-}H}$$

$$\mathrm{\xrightarrow{H_2O} C_6H_5{-}CH_2{-}C({=}O){-}H}$$

$$\mathrm{\xrightarrow{NH_3,\,HCN} C_6H_5{-}CH_2{-}CH(NH_2){-}C{\equiv}N \longrightarrow C_6H_5{-}CH_2{-}CH(NH_2){-}COOH}$$

Hydrogen sulfide can be added to phenylacetylene by a radical process rather than by the nucleophilic reaction described above. The HS radicals can be generated from an H_2S solution photochemically. The addition of HS radicals also leads to the hydroxylation of the phenyl ring of the phenylacetylene, by a mechanism which is not yet understood, to produce p-hydroxyphenylacetaldehyde. This molecule can undergo a Strecker reaction to give tyrosine.

Hydroxyl radicals would also react with phenylalanine to give tyrosine, along with ortho and meta hydroxyphenylalanine. The hydroxyl radicals required for tyrosine synthesis on the primitive earth might have been formed by the decomposition of hydrogen peroxide or the action of ionizing radiation on water.

SYNTHESIS FROM HYDROGEN CYANIDE

Dilute solutions of hydrogen cyanide undergo hydrolysis to formamide and formic acid.

$$\mathrm{HCN} \xrightarrow{H_2O} \mathrm{H{-}C({=}O){-}NH_2} \xrightarrow{H_2O} \mathrm{H{-}C({=}O){-}OH}$$

However, more concentrated solutions held at pH's fairly close to 9 undergo a complex sequence of polymerization reactions. The rate determining step seems to be the attack of CN^- on HCN. The dimer so formed then adds two further molecules of HCN to give the tetramer.

$$\mathrm{H{-}C{\equiv}N} + {}^{-}\mathrm{C{\equiv}N} \rightarrow \mathrm{H{-}C({=}N^{-}){-}C{\equiv}N} \xrightarrow{H^+} \mathrm{H{-}C({=}NH){-}C{\equiv}N}$$

$$\xrightarrow{HCN} \underset{\text{HCN trimer}}{\mathrm{N{\equiv}C{-}CH(NH^2){-}C{\equiv}N}} \xrightarrow{HCN} \underset{\text{HCN tetramer}}{\mathrm{(H_2N)(N{\equiv}C)C{=}C(C{\equiv}N)(NH_2)}}$$

The trimer of HCN has been prepared from malononitrile, but so far has not been isolated from a polymerizing solution of hydrogen cyanide. The tetramer, diaminomaleonitrile, is readily obtained in this way. Both the trimer and tetramer yield glycine on hydrolysis, but in concentrated

$$\mathrm{NC{-}CH(NH_2){-}CN} \rightarrow \mathrm{HOOC{-}CH(NH_2){-}COOH} \rightarrow \mathrm{CH_2(NH_2){-}COOH} + CO_2$$

solution it is clear that the tetramer is the more important intermediate. The formation of glycine from the tetramer may go through the water or ammonia adduct; for example,

$$\begin{matrix} H_2N & & C{\equiv}N \\ & C & \\ & \| & \\ & C & \\ H_2N & & C{\equiv}N \end{matrix} \xrightarrow{H_2O} \begin{matrix} & H & \\ H_2N & | & C{\equiv}N \\ & C & \\ & | & \\ & C & \\ H_2N & | & C{\equiv}N \\ & OH & \end{matrix} \longrightarrow \begin{matrix} H_2N{-}CH_2{-}C{\equiv}N \\ + \\ O \\ \| \\ H_2N{-}C{-}C{\equiv}N \end{matrix}$$

$$\longrightarrow \underset{\displaystyle NH_2}{\overset{|}{CH_2}}{-}COOH + CO_2 + HCOOH + 3NH_3$$

In addition to the well-defined substances mentioned above, solutions of hydrogen cyanide at moderately alkaline pH's deposit a very heterogeneous brown polymer. It seems that the tetramer initiates this further polymerization, but the mechanism is not understood. Hydrolysis, particularly of the more soluble polymer fractions, yields a series of α-amino acids. In addition to glycine and urea, among the natural amino acids, aspartic acid, alanine, serine, and probably threonine and glutamic have been identified. The polymer is so intractable that despite a considerable amount of effort, virtually nothing has been learned about the mechanism of this reaction. Claims that well-defined peptides are present in the polymers do not seem justified, although amide bonds of some type may be present.

Valine or isoleucine (as well as alloisoleucine) can be synthesized from a cyanide polymerization reaction if acetone or methyl-ethyl ketone is added. The synthesis is probably the result of the addition of the ketone to glycine nitrile (see above). In other experiments, these amino acids were obtained directly from the ketones and aminoacetonitrile.

$$CH_3{-}\overset{\displaystyle O}{\overset{\|}{C}}{-}CH_3 + \underset{\displaystyle NH_2}{\overset{|}{CH_2}}{-}C{\equiv}N \xrightarrow{-H_2O} \begin{matrix} CH_3 & \\ & C{=}\underset{\displaystyle NH_2}{\overset{|}{C}}{-}C{\equiv}N \\ CH_3 & \end{matrix}$$

$$\xrightarrow{\text{reduction}} \begin{matrix} CH_3 & \\ & CH{-}\underset{\displaystyle NH_2}{\overset{|}{CH}}{-}C{\equiv}N \\ CH_3 & \end{matrix} \xrightarrow{\text{hydrolysis}} \begin{matrix} CH_3 & \\ & CH{-}\underset{\displaystyle NH_2}{\overset{|}{C}}HCOOH \\ CH_3 & \end{matrix}$$

The nature of the reducing agent is not known. The direct synthesis of valine and isoleucine in the electric discharge was probably more important than this reaction sequence on the primitive earth.

OTHER AMINO ACID SYNTHESES

More complex amino acids can be obtained from simple ones by modification in aqueous solution. An interesting reaction of this type is the addition of aldehydes to polyglycine.

$$HCHO + \text{polyglycine} \rightarrow \text{polyserine}$$

$$CH_3CHO + \text{polyglycine} \rightarrow \text{polythreonine}$$

Formaldehyde will also add to free glycine in solution. The rate is constant above pH 9 and falls off at pH <8.

$$\underset{NH_3^+}{\underset{|}{CH_2}}-COO^- \underset{}{\overset{OH^-}{\rightleftarrows}} {}^-\underset{NH_3^+}{\underset{|}{CH}}-COO^- \overset{HCHO}{\rightleftarrows} \underset{OH}{\underset{|}{CH_2}}-\underset{NH_3^+}{\underset{|}{CH}}-COO^-$$

The reaction is reversible, but serine synthesis is favored at concentrations of HCHO > 10^{-3} M. The enzyme serine hydroxymethyl transferase catalyzes this reaction. The kinetics of the nonenzymatic reaction are rather complicated, so it is difficult to extrapolate to lower temperatures from experiments carried out at 60–100°C. Nevertheless, this reaction may have been a significant source of serine on the primitive earth.

Ultraviolet light catalyzes the addition of alkenes to polyglycine absorbed on bentonite.

$$CH_3CH{=}CH_2 + \text{polyglycine} \rightarrow \text{polyvaline}$$

A similar reaction with $CH_3CH{=}CHCH_3$ or $CH_3CH_2CH{=}CH_2$ gives leucine, isoleucine, and alloisoleucine. Few details about this reaction are known.

A plausible synthesis of tryptophan involves the addition of indole to a derivative of dehydroalanine.

$$\text{indole} + CH_2{=}\underset{HN-\underset{\|}{\overset{}{C}}(O)-CH_3}{\underset{|}{C}}-\overset{O}{\overset{\|}{C}}-NH_2 \longrightarrow \text{indol-3-yl}-CH_2-\underset{NH-\underset{\|}{C}(O)-CH_3}{\underset{|}{CH}}-\overset{O}{\overset{\|}{C}}-NH_2$$

Indole can be synthesized in substantial yield by the action of electric discharges or very high temperatures on mixtures of hydrocarbons with ammonia. The highest yields (3.1%) are obtained by the pyrolysis of a mixture of acetylene and NH_3.

The addition of indole to N-acetyl dehydroalanine amide or other dehydroalanine derivatives gives a yield of tryptophan of several percent. However, these reactions are relatively slow (even at 60°C). The rates at lower temperatures have not been measured, but they are probably high enough to have given small yields of tryptophan on the primitive earth.

Amino acids have been synthesized in very low yield by heating CH_4, NH_3, and H_2O to about 1,000°C in silica tubes. The major products of this pyrolysis are known to be hydrogen cyanide and acetylene. Thus, the function of the hot tube is probably to provide an alternative source of cyanide and aldehydes as intermediates in a Strecker-type synthesis.

A variety of intraconversions between amino acids in aqueous solution are brought about by ultraviolet or high energy radiation. An example is

$$CH_3-\underset{\displaystyle NH_2}{\underset{|}{CH}}-COOH + CO_2 \xrightarrow{\gamma \text{ ray}} \text{aspartic acid}$$

In general, the yields are low and the product mixtures complex. It is doubtful whether these reactions were important for prebiotic synthesis since they require very high radiation intensities, usually high enough to destroy the products in a relatively short time.

In summary, many, but not all, of the twenty amino acids that occur in proteins have been synthesized under primitive earth conditions. Asparagine and glutamine, the amides of aspartic and glutamic acid, are synthesized as intermediates during the hydrolysis of the corresponding nitriles, but are too labile to permit easy isolation. There are no convincing syntheses for arginine, lysine, and histidine, and some of the other syntheses are very inefficient. It should be emphasized that many amino acids that do not occur in proteins are obtained in some of these prebiotic syntheses—for example, sarcosine, α-aminobutyric acid, and norvaline. The yields of these nonprotein amino acids are comparable to those of their naturally occurring isomers.

UREA

In 1828, Wöhler obtained urea in high yield by heating ammonium cyanate. This was the first synthesis of an organic compound from inorganic starting materials, although it was not immediately recognized as such. As we shall see, this is also an acceptable prebiotic synthesis.

More recently, urea has been obtained in a number of studies in prebiotic chemistry. It is obtained when an electric discharge is passed through a mixture of methane, ammonia, and water. It is also formed when the products from the polymerization of hydrogen cyanide are hydrolyzed. Excellent yields are obtained when an ammonium cyanide solution is irradiated in the presence of ferrous iron.

The cyanate ion is readily formed by the hydrolysis of cyanogen in neutral or alkaline solutions.

$$H_2O + N{\equiv}C{-}C{\equiv}N \rightarrow H_2N{-}\overset{\overset{\displaystyle O}{\|}}{C}{-}C{\equiv}N$$

$$\rightarrow {}^{-}N{=}C{=}O + CN^{-} + 2H^{+}$$

The rate of formation of urea from NH_4^+ and cyanate is independent of pH between pH's of 5 and 9, because the reactive species are ammonia and cyanic acid.

$$HN{=}C{=}O + NH_3 \rightarrow \left[{}^{-}HN{-}\overset{\overset{\displaystyle O}{\|}}{C}{-}NH_3^+\right] \rightarrow H_2N{-}\overset{\overset{\displaystyle O}{\|}}{C}{-}NH_2$$

The equilibrium constant at 25°C for the reaction

$$NH_4^+ + {}^{-}N{=}C{=}O \rightleftharpoons H_2N{-}\overset{\overset{\displaystyle O}{\|}}{C}{-}NH_2$$

is estimated to be 3×10^4.

One mechanism of hydrolysis of urea is the reversal of the Wöhler synthesis followed by hydrolysis of the cyanate to carbamate. Carbamate is hydrolyzed in turn to ammonia and carbon dioxide.

$${}^{-}N{=}C{=}O + H_2O \rightarrow H_2N{-}\overset{\overset{\displaystyle O}{\|}}{C}{-}O^{-} \xrightarrow{H^+} NH_3 + CO_2$$

It is clear that urea could have been formed in large amounts on the primitive earth. The half-life for the decomposition of urea to cyanate and NH_4^+ is 6,300 years at 0°C and 41 years at 25°C. On the primitive earth, some of the cyanate would have reacted with NH_4^+ to reform urea, and some would have hydrolyzed to ammonia and carbon dioxide. As long as the temperature was low enough, considerable amounts of urea could have accumulated.

SYNTHESIS ON MINERAL SURFACES

A novel approach to prebiotic synthesis has been reported by Horowitz, Hubbard, and Hardy. The Martian atmosphere contains mostly CO_2 at a pressure of about 6×10^{-3} atm. The CO is about 0.1% of the CO_2, and water vapor is very low, about 5×10^{-7} atm. When glass beads in contact with such an atmosphere were irradiated with ultraviolet light, good yields of formaldehyde and glycolic acid were obtained.

This reaction is particularly interesting because it may take place on the surface of Mars or on dust grains in interstellar clouds at the present time. The yields are good even when light of wavelength as long as 3,200 Å is used. There is no reaction at all in the gas phase at this wavelength. In addition, this is one of the few prebiotic syntheses that proceeds in an atmosphere which is only moderately reducing. (It is reducing because it contains CO.)

It will be interesting to extend these experiments to gas mixtures containing nitrogen and ammonia. It is possible that many of the reactions which have so far been carried out only under strongly reducing conditions will proceed under the conditions used by Horowitz. If so, we may have to revise many of our ideas about the conditions under which prebiotic syntheses are possible. In any case, this synthesis may have been important for the synthesis of formaldehyde, which in turn could have condensed to give sugars.

FISCHER–TROPSCH SYNTHESES

The Fischer–Tropsch reaction is used to make hydrocarbons from carbon monoxide and hydrogen—e.g., octane.

$$8CO + 17H_2 \xrightarrow{\text{catalyst}} n\text{—}C_8H_{18} + 8H_2O$$

The reaction is usually carried out at pressures between 1 and 50 atm and at temperatures ranging from 180° and 300°C. A wide variety of catalysts can be used; a typical one is a mixture of iron and nickel supported on silica. It is interesting that straight chain hydrocarbons are obtained as final products provided that the contact time in the catalyst bed is not so long that the straight chain hydrocarbons, which are formed initially, undergo isomerization.

Anders and his colleagues have shown that a wide variety of aliphatic and aromatic hydrocarbons, alcohols, etc. can be made by this reaction. When a mixture of CO, H_2, and NH_3 is used, amino acids, purines, pyrimidines, and a wide variety of other compounds are produced. The yields of these products are relatively low, and are greatly dependent on the experimental conditions. Some experiments are performed by passing the gas mixture once over the catalyst and quenching the reaction; other experiments allow the gas to be in continuous contact with the catalyst but have a cold area in the vessel to "freeze out" the high boiling materials. The mechanisms of synthesis of the products are not clear, but since hydrogen cyanide is a major product of the $CO + H_2 + NH_3$ reaction, the mechanisms may be similar to those outlined for other prebiotic amino acid, purine, and pyrimidine syntheses.

It is proposed that this type of synthesis occurred in the solar nebula and is responsible for the organic compounds found in the Carbonaceous Chondrites (Chapter 15). It is also proposed that such reactions took place on the primitive earth. There are difficulties in carrying out a Fischer–Tropsch reaction under geological conditions. One problem is that sulfur compounds poison Fischer–Tropsch catalysts. Another problem is to obtain the correct contact time under geological conditions. If the contact time is too long, then an equilibrium mixture of CH_4, H_2, H_2O, and CO will result rather than a mixture of higher hydrocarbons and other interesting products. There is no great problem in producing a mixture of $CO + H_2$ since the equilibrium for the reaction

$$CH_4 + H_2O \underset{\text{low T}}{\overset{\text{high T}}{\rightleftharpoons}} CO + 3H_2$$

is toward the right at high temperatures. On the other hand, high temperatures decompose NH_3 to N_2 and H_2, so mixtures of $CO + H_2 + NH_3$ do not seem to be geologically reasonable unless some complicated mixing process is assumed. Thus we doubt that the Fischer–Tropsch reaction is as important as electric discharge or photochemical reactions for organic compound synthesis on the primitive earth (or in the Carbonaceous Chondrites). However, it may have played a significant role in the synthesis of some types of organic compounds.

FATTY ACIDS

No satisfactory synthesis of fatty acids is at present available. The action of electric discharges on methane and water gives fairly good yields of acetic and propionic acids, but only small yields of the higher fatty acids. Furthermore, the small quantities of higher fatty acids that are formed are highly branched.

It should not be too difficult to synthesize straight chain fatty acids from ethylene under prebiotic conditions. The irradiation with γ rays of ethylene and CO_2 at high pressures produces very long chain fatty acids. However, no variation of this type of experiment which yields fatty acids under primitive earth conditions has yet been demonstrated.

PORPHYRINS

Although porphyrins are not essential for life in the sense that nucleic acids are, they may have played a significant role in prebiotic chemistry. Almost certainly they became important for the metabolic processes leading to ATP synthesis early in the evolution of life.

A synthesis of porphyrins, which is proposed to be prebiotic, is based on the reaction of formaldehyde with pyrrole.

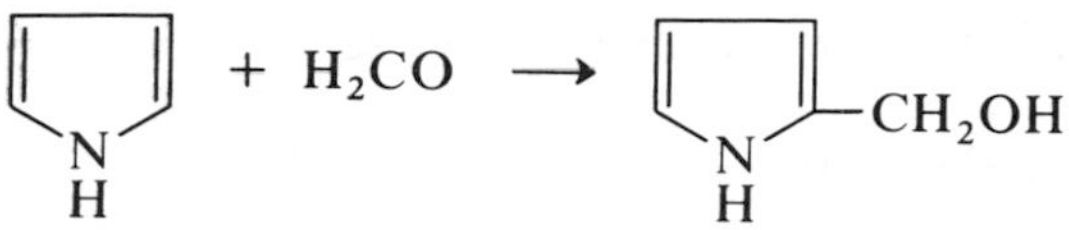

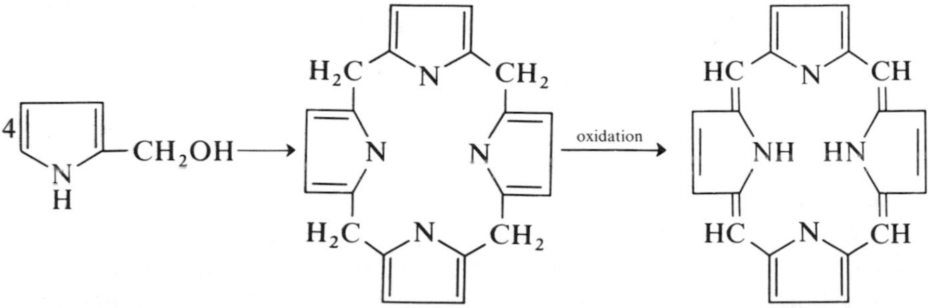

$$+\ H_2CO \longrightarrow \quad -CH_2OH$$

$$4\ \ -CH_2OH \longrightarrow \quad \xrightarrow{\text{oxidation}}$$

This reaction takes place under relatively dilute conditions and is catalyzed somewhat by metal ions that can chelate the pyrrole nitrogens. The action of electric discharge on a mixture of CH_4, NH_3, and H_2O gives excellent yields of formaldehyde and small yields of pyrrole. Very small yields of porphyrins have been claimed from such electric discharge reactions.

It seems likely that there are more efficient prebiotic syntheses of porphyrins, especially of those that are closer in structure to the protoporphyrins and uroporphyrins that are used in present day biological systems.

NICOTINAMIDE AND ITS DERIVATIVES

Nicotinamide is the only vitamin for which prebiotic syntheses have been proposed. The simplest route is via nicotinonitrile, which is obtained in 10^{-3}% yield by the action of an electric discharge on a mixture of ethylene and ammonia. It is believed that pyridine is formed as an intermediate and reacts with cyanide in the discharge. Pyridine can be formed in the discharge in yields up to 0.1%, and has been shown in independent experiments to react with hydrogen cyanide under the conditions of the experiment to give a mixture of 2-, 3-, and 4-cyanopyridines.

An alternative synthesis makes use of cyanoacetylene, propiolaldehyde, and ammonia. We have already described the synthesis of cyanoacetylene from nitrogen and methane. A mixture of methane, nitrogen, and water, under similar conditions, produced propiolaldehyde in up to 0.5% yield, probably by the hydration of C_3.

$$C_3 + H_2O \rightarrow HC{\equiv}C{-}C({=}O){-}H$$

Nicotinonitrile is probably obtained via the sequence

$$HC{\equiv}C{-}C{\equiv}N \rightarrow OHC{-}CH_2{-}C{\equiv}N$$

$$\xrightarrow{HC\equiv C-CHO} \quad HC(=O)-HC{=}CH-CH(-C{\equiv}N)-CH{=}O \quad \xrightarrow{NH_3} \quad \text{(3-cyanopyridine ring)}{-}C{\equiv}N$$

Nicotinonitrile is readily hydrolyzed to nicotinamide (Vitamin B_3).

Direct synthesis in an electric discharge has the merit of simplicity, but the yields are small and a mixture of isomeric cyanopyridines is obtained. The synthesis from propiolaldehyde gives better yields of nicotinonitrile without the other isomers. However, it does depend on a relatively complex series of reactions. It is hard to estimate the relative contributions of the two reactions to the synthesis of nicotinonitrile on the primitive earth.

References

There have been many attempts, which were largely unsuccessful, to synthesize organic compounds nonbiologically under oxidizing conditions (that is, from CO_2 and H_2O). Most of these experiments were intended to be models of plant photosynthesis, but some were done as prebiotic experiments. A review and two papers are:

E. I. Rabinowitch, *Photosynthesis* (Interscience, New York, 1945). Vol. I, pp. 61–98.

W. Groth and H. Suess, "Bemerkungen zur Photochemie der Erdatmosphäre," *Naturwissenschaften* **26**, 77 (1938).

W. M. Garrison, D. C. Morrison, J. G. Hamilton, A. A. Benson, and M. Calvin, "Reduction of Carbon Dioxide in Aqueous Solutions by Ionizing Radiation," *Science* **114**, 416 (1951).

The electric discharge synthesis of amino acids was first reported in 1953, with details appearing subsequently:

S. L. Miller, "A production of amino acids under possible primitive earth conditions," *Science* **117**, 528 (1953).

S. L. Miller, "Production of some organic compounds under possible primitive earth conditions," *J. Amer. Chem. Soc.* **77**, 2351 (1955).

S. L. Miller, "The mechanism of synthesis of amino acids by electric discharges," *Biochim. Biophys. Acta* **23**, 480 (1957).

S. L. Miller, "The formation of organic compounds on the primitive earth," *Ann. N.Y. Acad. Sci.* **69**, 260 (1957); *The Origin of Life on the Earth*, A. I. Oparin, ed. (Pergamon Press, Oxford, 1959), pp. 123–135.

The more recent syntheses using CH_4, N_2, H_2O, with traces of NH_3 are:

D. Ring, Y. Wolman, N. Friedmann, and S. L. Miller, "Prebiotic synthesis of hydrophobic and protein amino acids," *Proc. Nat. Acad. Sci.* **69**, 765 (1972).

Y. Wolman, W. H. Haverland, and S. L. Miller, "Non-protein amino acids from spark discharges and their comparison with the Murchison meteorite amino acids," *Proc. Nat. Acad. Sci.* **69**, 809 (1972).

Syntheses of amino acids by ultraviolet light from the reducing atmosphere constituents:

W. Groth and H. v. Weyssenhoff, "Photochemical formation of organic compounds from mixtures of simple gases," *Planet. Space Sci.* **2**, 79 (1960).

C. Sagan and B. N. Khare, "Long-wavelength ultraviolet photoreproduction of amino acids on the primitive earth," *Science* **173**, 417 (1971); *Nature*, **232**, 577 (1971).

Synthesis of amino acids by shock waves:

A. Bar-Nun, N. Bar-Nun, S. H. Bauer, and C. Sagan, *Science* **168**, 470 (1970).

The synthesis of amino acids from NH_4CN polymerization was first reported by Oró in 1961:

J. Oró and S. S. Kamat, "Amino acid synthesis from hydrogen cyanide under possible primitive earth conditions," *Nature* **190**, 442 (1961).

Other papers on amino acids from NH_4CN include:

C. U. Lowe, M. W. Rees, and R. Markham, *Nature* **199**, 219 (1963).

C. N. Matthews and R. E. Moser, *Proc. Nat. Acad. Sci. U.S.* **56**, 1087 (1966); *Nature* **215**, 1230 (1967).

R. E. Moser and C. N. Matthews, *Experentia* **24**, 658 (1968).

M. Labadie, R. Jensen, and E. Neuzil, *Bull. Soc. Chim. Biol.* **49**, 673 (1967); *Biochim. Biophys. Acta* **165**, 525 (1968).

N. Friedmann and S. L. Miller, *Nature* **221**, 1152 (1969).

The direct synthesis of the amino nitriles from methane and ammonia in the electric discharge is reported by C. Ponnamperuma and F. Woeller, *Curr. Mod. Biol.* **1**, 156 (1967).

Aspartic acid from cyanoacetylene was reported by R. A. Sanchez, J. P. Ferris, and L. E. Orgel, *Science* **154**, 784 (1966).

Phenylalanine and tyrosine from phenylacetylene were reported by N. Friedmann and S. L. Miller, *Science* **166**, 766 (1969).

Spark discharge synthesis of methionine and amino acid syntheses from acrolein are reported by J. E. Van Trump and S. L. Miller, *Science* **178**, 859 (1972).

Synthesis of serine, threonine, valine, isoleucine, and leucine by additions to polyglycine was reported by S. Akabori in *The Origin of Life on the Earth*, A. I. Oparin, *et al.*, eds. (Pergamon Press, Oxford, 1959), pp. 189–196.

Fatty acids (C_2 to C_{12})

W. V. Allen and C. Ponnamperuma, *Currents Mod. Biol.* **1**, 24 (1967).

Organic compounds from CO and H_2O

J. S. Hubbard, J. P. Hardy, and N. H. Horowitz, "Photocatalytic production of organic compounds in a simulated Martian atmosphere," *Proc. Nat. Acad. Sci.* **68**, 574 (1971).

Fischer–Tropsch syntheses

M. H. Studier, R. Hayatsu, and E. Anders, "Origin of organic matter in early solar system. I. Hydrocarbons," *Geochim. Cosmochim. Acta* **32**, 151 (1968).

R. Hayatsu, M. H. Studier, A. Oda, K. Fuse, and E. Anders. II. "Nitrogen compounds." *Geochim. Cosmochim. Acta* **32**, 175 (1968).

D. Yoshino, R. Hayatsu, and E. Anders. III. "Amino acids: catalytic synthesis," *Geochim. Cosmochim. Acta* **35**, 927 (1971).

R. Hayatsu, M. H. Studier, and E. Anders. IV. "Amino acids: Confirmation of catalytic synthesis by mass spectrometry," *Geochim. Cosmochim. Acta* **35**, 939 (1971).

M. H. Studier, R. Hayatsu, and E. Anders. V. "Further studies of meteoritic hydrocarbons and a discussion of their origin," *Geochim. Cosmochim. Acta* **36**, 189 (1972).

R. Hayatsu *et al.* VI. "Catalytic synthesis of nitriles, nitrogen bases and porphyrin-like pigments," *Geochim. Cosmochim. Acta* **36**, 555 (1972).

Porphyrins from pyrrole and formaldehyde and from the electric discharge

P. Rothemund, *J. Am. Chem. Soc.* **58**, 625 (1936).

A. A. Krasnovskii and A. V. Umrikhina, *Dokl. Akad. Nauk SSSR* **155**, 691 (1964); *Chem. Abst.* **61**, 858 (1964).

G. W. Hodgson and B. L. Baker, *Nature* **216**, 29 (1967).

G. W. Hodgson and C. Ponnamperuma, *Proc. Nat. Acad. Sci.* **59**, 22 (1968).

Nicotinonitrile and nicotinamide

M. J. Dowler, W. D. Fuller, L. E. Orgel, and R. A. Sanchez, *Science* **169**, 1320 (1970).

N. Friedmann, S. L. Miller, and R. A. Sanchez, *Science* **171**, 1026 (1971).

Chapter Eight

Prebiotic Synthesis of Purines, Pyrimidines, and Nucleosides

INTRODUCTION

In the last chapter we discussed the prebiotic synthesis of amino acids and certain other small molecules. We describe here the prebiotic synthesis of the organic components of nucleic acids, discussing in turn the synthesis of purines, pyrimidines, and sugars. Then we discuss the various routes to the nucleosides.

SYNTHESIS OF PURINES

The most important purines, from our point of view, are adenine **(1)** and guanine **(2)**.

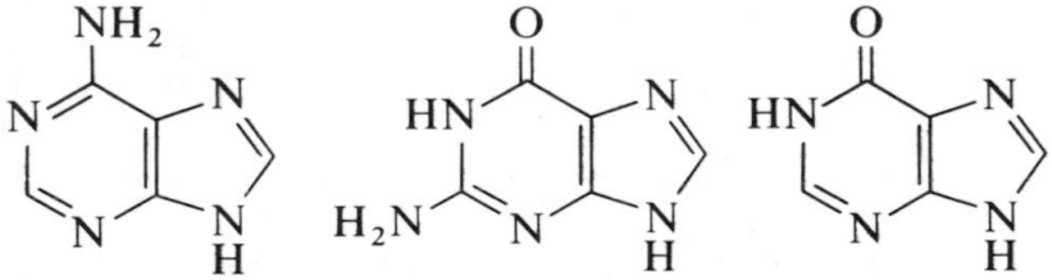

1 **2** **3**

These compounds are major constituents both of DNA and RNA and are therefore essential for the functioning of the genetic apparatus. In addition, hypoxanthine **(3)** is present in small amounts in certain t-RNAs; according to the "wobble" hypothesis it forms part of the anticodon in those t-RNAs which can recognize sets of three codons, XYU, XYC, XYA (Chapter 6). Adenine is also a constituent of ATP, the major intermediate in the utilization of metabolic energy.

The central role of adenine, both in genetic processes and in energy utilization, makes all the more striking the remarkable ease with which it is synthesized from ammonium cyanide. Oró and his coworkers discovered that if concentrated ammonia solutions containing cyanide (1–10 M) are refluxed, adenine is formed in yields of up to 1/2%, along with large amounts of the usual hydrogen cyanide polymer. The reaction conditions are not critical, for adenine has subsequently been obtained from concentrated aqueous ammonium cyanide solutions under a variety of conditions.

It should be emphasized that the formation of a molecule as complex as adenine from cyanide, even though in only 1/2% yield, is quite unexpected. Even more so is the commercially useful synthesis of adenine in up to 30% yield from hydrogen cyanide in liquid ammonia. Since hydrogen cyanide is readily obtained under prebiotic conditions, and since it is an intermediate in most of the plausible prebiotic syntheses of amino acids, the finding that it is also a potential source of adenine has greatly impressed most students of the origins of life.

Oró's studies, and those of others, have thrown much light on the mechanism of purine synthesis from cyanide. On the one hand, hydrogen cyanide undergoes hydrolysis to formamide and formic acid, or aminolysis to formamidine.

$$HCN \xrightarrow{H_2O} H{-}\overset{\overset{\displaystyle O}{\|}}{C}{-}NH_2 \xrightarrow{H_2O} HCOO^- + NH_4^+$$

$$HCN + NH_3 \rightarrow H_2N{-}CH{=}NH$$

On the other hand, it undergoes self-condensation to give the important tetramer of hydrogen cyanide, diaminomaleonitrile

$$H{-}C{\equiv}N + {}^-C{\equiv}N \rightarrow H{-}\overset{\overset{\displaystyle NH}{\|}}{C}{-}C{\equiv}N \rightarrow N{\equiv}C{-}\overset{\overset{\displaystyle NH_2}{|}}{CH}{-}C{\equiv}N$$

Aminomalonitrile
(HCN trimer)

$$(H_2N)(N{\equiv}C)CH{-}C{\equiv}N + {}^-C{\equiv}N \xrightarrow{H^+} (H_2N)(N{\equiv}C)C{=}C(C{\equiv}N)(NH_2)$$

Diaminomaleonitrile
(HCN tetramer)

The tetramer plays a central role in cyanide chemistry. We have already seen that in aqueous cyanide solution it may undergo hydrolysis to give glycine, or it can initiate the polymerization of further cyanide molecules to give dark-colored solids. In ammoniacal hydrogen cyanide, it may also react with formamidine formed *in situ* to give either 4-aminoimidazole-5-carbonitrile **(4)** or the corresponding amidine **(5)**.

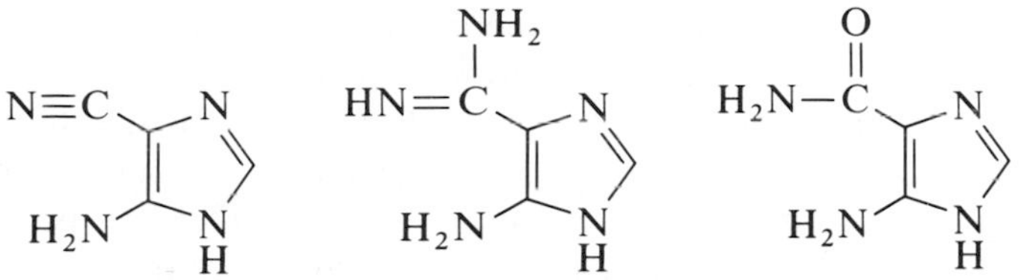

4 **5** **6**

Either of these compounds will react with hydrogen cyanide or formamidine to give adenine.

In addition to the compounds discussed above, Oró isolated 4-amino-imidazole-5-carboxamide **(6)**, the hydrolysis product of the carbonitrile and carboxamidine. This substance, which is an intermediate (as the riboside) in the biosynthesis of purines, turns out to be a key intermediate in potentially prebiotic syntheses of guanine and hypoxanthine.

The chemistry described above does not easily explain the synthesis of purines under prebiotic conditions because useful yields of adenine cannot be obtained except in the presence of 1.0 M or stronger ammonia. The highest reasonable concentration of ammonia or ammonium ion that can be postulated in oceans and lakes on the primitive earth is about 0.01 M.

It should be clear that ammonia is essential in only one step of the adenine synthesis outlined above, namely, the formation of the formamidine. This is subsequently used to convert tetramer to **(4)** and **(5)**. This step can be avoided by making use of an unusual photochemical rearrangement, the conversion of tetramer to **(4)**.

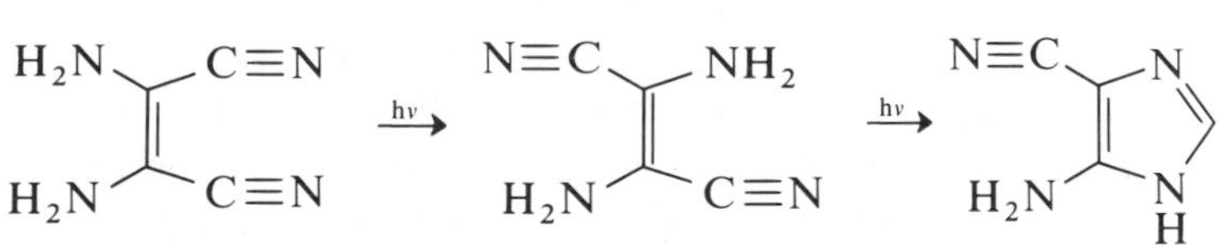

4

This reaction proceeds with high yield in sunlight. We can thus describe a synthesis of adenine from aqueous hydrogen cyanide which requires no additional reagents

$$4\text{HCN} \rightarrow \text{tetramer} \xrightarrow{h\nu} 4 \xrightarrow{\text{HCN}} \text{adenine}$$

We must next ask about the physical conditions which favor this reaction. The rate of tetramer formation is maximal at the pK_a of hydrogen cyanide, that is at pH = 9.2 at room temperature. It proceeds at a reasonable rate in the "prebiotic" pH range of 7–10. The reaction is rapid at 100°C but proceeds at a measurable rate at all temperatures which have been studied, down to −20°C. Thus, neither the temperature nor the pH is likely to have been an obstacle to tetramer synthesis on the primitive earth. More limited studies of the conversion of (4) to adenine suggest that it too could occur under a wide range of prebiotic conditions.

When we come to consider the concentration dependence, however, we run into a serious obstacle. Hydrolysis of cyanide to formamide and formic acid becomes the dominant mode of reaction of cyanide once its concentration falls below about 0.01 M. Thus, yields of adenine must have been very low unless some way of building up high cyanide concentration was available. Hydrogen cyanide cannot be concentrated by evaporation since it is too volatile, so we must either postulate a mechanism of generating a strong solution initially or provide an alternative concentration mechanism.

The rapid hydrolysis of cyanide to formate would not permit a slow accumulation of cyanide from the atmosphere over geological time periods. High concentrations could be achieved close to localized sources (e.g., in raindrops close to thunderstorms) if cyanide was formed from methane and nitrogen in the atmosphere. Alternately, cyanide might possibly have accumulated in the atmosphere prior to the formation of the oceans and then dissolved in the nascent oceans as soon as they began to form.

We believe that a much more plausible way of explaining the formation of concentrated cyanide solutions on the primitive earth makes use of *eutectic* freezing. If dilute cyanide solutions are cooled below 0°C, ice separates out and all the hydrogen cyanide remains in the liquid phase until at −22°C, the whole mass solidifies. The liquid phase at the eutectic temperature contains about 75% of hydrogen cyanide. In the range 0° to −20°C, the concentration of the cyanide solution in equilibrium with ice is, of course, independent of the initial concentration, so that enormous enrichments are possible (see Chapter 10).

It has been shown that good yields of tetramer can be obtained from cyanide solutions as dilute as 0.001 M at −10°C. In addition, the usual polymer, which is a potential source of amino acids, is formed. In free solution above 0°C, negligible amounts of biologically interesting material would be obtained from such dilute solutions because all the cyanide would hydrolyze to formic acid.

Instead of freezing a dilute solution of cyanide, it is possible in the laboratory to start with ice beneath an atmosphere containing a small amount of hydrogen cyanide. The hydrogen cyanide dissolves to form a surface layer of eutectic solution, provided that its vapor pressure exceeds that of the cyanide solution in equilibrium with ice at the working temperature. From our point of view, this provides a model for an attractive method of generating concentrated cyanide solutions under conditions which would automatically expose them to sunlight.

Finally, we must note that the presence of large amounts of salts greatly lowers the efficiency of cyanide polymerization in eutectics, since the eutectic volume is then determined by the amount of salt present rather than the cyanide. The concentration which can be effected is thus reduced. This makes the process less likely in the oceans, although it could occur in lakes (or frozen raindrops). Of course, the frozen surface of lakes or oceans would be equally suitable sites for purine synthesis.

The oxidation state of guanine is higher than that of adenine so that its formation from 4-aminoimidazole-5-carbonitrile derivatives requires the addition of a one-carbon molecule at the oxidation level of CO_2. Also, in order to obtain an OH group rather than an NH_2 group at C_6, we must start with the carboxamide rather than the corresponding nitrile or amidine.

The synthesis of guanine has not been studied in such detail as the corresponding synthesis of adenine. However, it is known that guanine can be obtained from **(4)** and cyanate, urea, or cyanogen. The reaction of choice, that with cyanogen, will be described here.

The introduction of cyanogen into a solution of **(4)** at room temperature leads to the rapid formation of a brown precipitate and to the removal of the starting material from solution. The brown solid then hydrolyzes, slowly at room temperature and pH 7, or more rapidly at higher temperatures or higher pH's, to give a well-defined intermediate, perhaps **(7a)** or **(7b)**. The

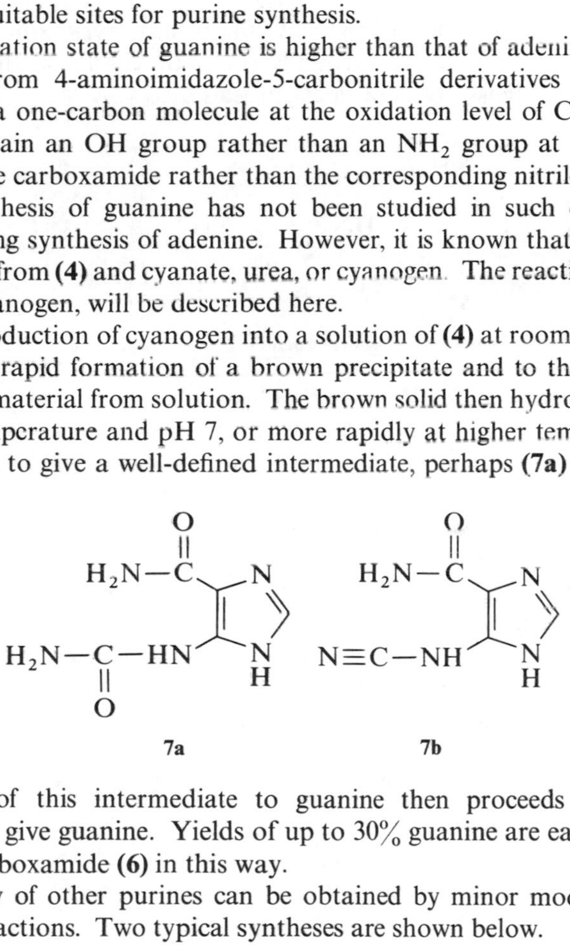

7a **7b**

cyclization of this intermediate to guanine then proceeds cleanly and efficiently to give guanine. Yields of up to 30% guanine are easily obtained from the carboxamide **(6)** in this way.

A variety of other purines can be obtained by minor modifications of the above reactions. Two typical syntheses are shown below.

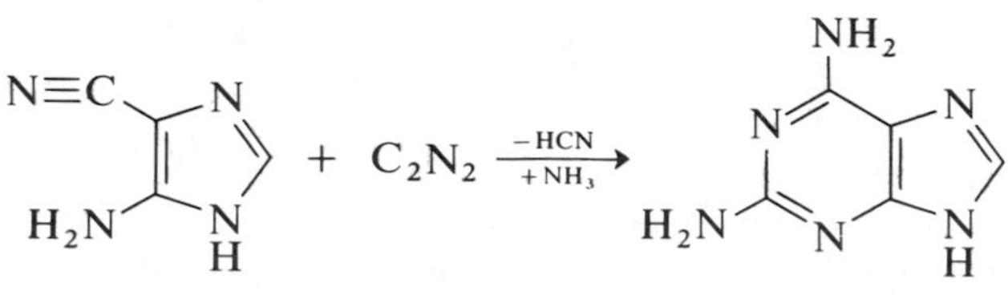

$$+ \; C_2N_2 \xrightarrow[+NH_3]{-HCN}$$

Diamino purine

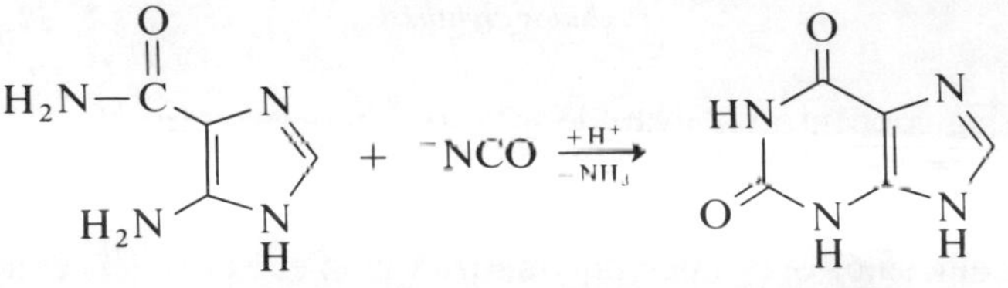

Xanthine

SYNTHESIS OF PYRIMIDINES

The synthesis of the biologically important pyrimidines—cytosine, uracil, and thymine—under potentially prebiotic conditions has not been investigated as extensively as purine synthesis.

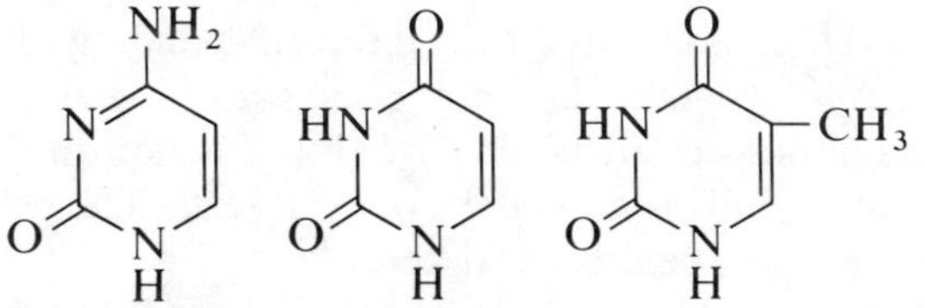

Cytosine **Uracil** **Thymine**

A study of the products of the action of an electric discharge on a mixture of methane and nitrogen led to the identification of cyanoacetylene. In fact, it was the major nitrogen-containing compound after hydrogen cyanide. More recently it has become clear that cyanoacetylene is also produced from the reaction of nitrogen atoms with acetylene, and when hydrocarbons and nitrogen are heated to temperatures above 700°C. Thus, the "high-energy" experiments which produce hydrogen cyanide also give a considerable amount of cyanoacetylene.

It was found that cyanoacetylene reacts directly with aqueous cyanate to give cytosine in quite good yield (20%), but that the more obvious reaction of cyanate with aminoacrylonitrile does not work.

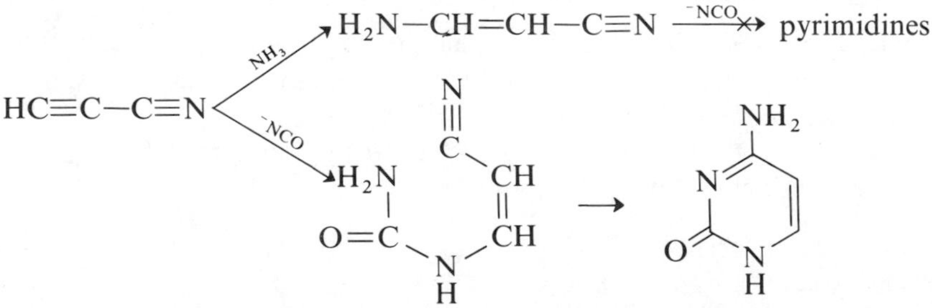

$H_2N{-}CH{=}CH{-}C{\equiv}N \xrightarrow{^-NCO} \not\to$ pyrimidines

$HC{\equiv}C{-}C{\equiv}N$

8 **Cytosine**

The mechanism of this synthesis is still not entirely clear. Cytosine does indeed arise by ring closure from an aminoacrylonitrile derivative, ureido-acrylonitrile **(8)**.

This compound has been isolated from solutions of cyanate and cyanoacetylene, either after warming or after standing for a few hours at room temperature. On heating, or at room temperature (but more slowly), ring-closure occurs to give cytosine in almost 100% yield. We do not know the sequence of steps leading to **(8)**, but the following is quite plausible.

$$HC{\equiv}C{-}C{\equiv}N \xrightarrow{^{-}NCO}$$

$$O{=}C{=}N{-}CH{=}CH{-}C{\equiv}N \xrightarrow{^{-}NCO} O{=}C(N{=}C{=}O){-}NH{-}CH{=}CH{-}C{\equiv}N \longrightarrow O{=}C(NH_2){-}NH{-}CH{=}CH{-}C{\equiv}N \;(\mathbf{8}) + CO_2$$

A second, closely related route to cytosine has also been found, namely, the direct reaction of aminoacrylonitrile with cyanogen or cyanamide. This will be important in our later discussion of nucleoside synthesis.

The sequence of reactions described above gives reasonable yields of cytosine, but only if the concentrations of the reactants exceed about 0.01 M. Such a concentration could not have been built up by slow accumulation in an aqueous phase since both cyanate and cyanoacetylene undergo hydrolysis to inactive products in at most a few hundred years. Perhaps eutectic concentration would be helpful here as in purine synthesis, but it has not been attempted as yet.

Uracil is obtained in very good yield by the hydrolysis of cytosine. The half-life varies from 19 days at 100°C to 200 years at 30°C. Thus, provided cytosine can be made, it is quite easy to obtain uracil from it. The direct synthesis of uracil from propiolamide ($HC{\equiv}C{-}CONH_2$) parallels the reaction from cyanoacetylene, but it is unlikely to have prebiotic significance since the hydrolysis of cyanoacetylene gives cyanoacetaldehyde and little or no propiolamide.

SUGARS

Formaldehyde is one of the simplest organic molecules and is readily formed under prebiotic conditions. The observation that sugars are formed by treating it with certain alkaline catalysts goes back to the work of Butlerov in the mid-nineteenth century. More recently, this synthesis, called the *formose reaction*, has been studied in greater detail using modern chromatographic techniques and has been discussed extensively as a model for prebiotic synthesis.

HOCH$_2$ O OH OH OH

β-D-ribose

HOCH$_2$ O OH HO OH

β-D-arabinose

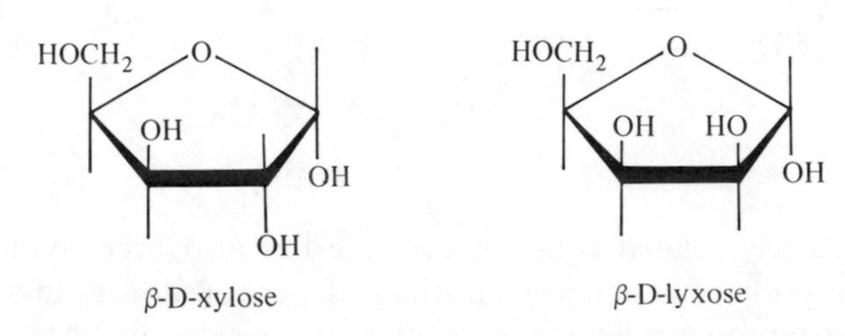

Fig. 8-1. The β-D-pentoses. Arabinose is derived from ribose by inversion of the 2-hydroxyl, xylose by inversion of the 3-hydroxyl and lyxose by inversion of both the 2- and the 3-hydroxyl. Lyxose is rare in biological systems. All these pentoses are formed in varying amounts in the formose reaction.

The most important aspects of the reaction may be summarized as follows:

1. Treatment of formaldehyde with a strong alkali, such as sodium hydroxide, brings about the Cannizzaro reaction

$$2HCHO \xrightarrow{OH^-} CH_3OH + HCO_2^-$$

but does not give much sugar. However, use of thallium hydroxide or lead hydroxide gives mainly sugars. Alternatively, the formose reaction may be brought about heterogeneously, using, for example, calcium carbonate, barium hydroxide, or alumina. In most published work, alkaline conditions have been used, but this is not essential with all catalysts. Alumina, for example, is quite a good catalyst at pH 7.4.

2. The reaction is autocatalytic. The induction period may last for many hours under conditions in which the reaction, once initiated, is over rapidly.

3. The earliest product identified is glycolaldehyde. Subsequently, glyceraldehyde and dihydroxyacetone appear, followed by tetroses, pentoses, and hexoses including ribose, and glucose, for example. Yields of up

to 50% of glycolaldehyde or higher sugars can be obtained if the reaction is stopped at the appropriate time.

4. The induction period can be eliminated by the addition of one of a variety of aldehydes and ketones—for example, glycolaldehyde, glyceraldehyde, or dihydroxyacetone. The bottleneck responsible for the induction period must therefore be the formation of glycolaldehyde from two molecules of formaldehyde.

$$2HCHO \rightarrow \underset{OH}{\underset{|}{CH_2}}-CHO \tag{1}$$

Since the reaction is autocatalytic, there must be a step in which one intermediate splits to give two reactive molecules. The simplest reaction scheme which fits the available evidence is

$$HCHO + \underset{OH}{\underset{|}{CH_2}}-CHO \rightarrow \underset{OH}{\underset{|}{CH_2}}-\underset{OH}{\underset{|}{CH}}-CHO \tag{2}$$

$$\underset{OH}{\underset{|}{CH_2}}-\underset{OH}{\underset{|}{CH}}-CHO \rightleftarrows \underset{OH}{\underset{|}{CH_2}}-\overset{O}{\overset{\|}{C}}-\underset{OH}{\underset{|}{CH_2}} \tag{3}$$

$$HCHO + \underset{OH}{\underset{|}{CH_2}}-\overset{O}{\overset{\|}{C}}-\underset{OH}{\underset{|}{CH_2}} \rightarrow \underset{OH}{\underset{|}{CH_2}}-\underset{OH}{\underset{|}{CH}}-\overset{O}{\overset{\|}{C}}-\underset{OH}{\underset{|}{CH_2}} \tag{4}$$

$$\underset{OH}{\underset{|}{CH_2}}-\underset{OH}{\underset{|}{CH}}-\overset{O}{\overset{\|}{C}}-\underset{OH}{\underset{|}{CH_2}} \rightleftarrows \underset{OH}{\underset{|}{CH_2}}-\underset{OH}{\underset{|}{CH}}-\underset{OH}{\underset{|}{CH}}-CHO \tag{5}$$

$$\underset{OH}{\underset{|}{CH_2}}-\underset{OH}{\underset{|}{CH}}-\underset{OH}{\underset{|}{CH}}-CHO \rightleftarrows 2\underset{OH}{\underset{|}{CH_2}}-CHO \tag{6}$$

Condensation of glycolaldehyde, trioses, etc. to pentose and hexose sugars, e.g.

$$\text{glyceraldehyde} + \text{glycolaldehyde} \rightarrow \text{ribose} \tag{7}$$

All of the steps suggested above, except (1), are analogues of well-known organic reactions. Reactions (2), (4), and (7) are aldol condensations, while (6) is a reverse aldol reaction. Reactions (3) and (5) are isomerizations of α-hydroxy aldehydes to α-hydroxy ketones.

$$\underset{\substack{|\\ \text{OH}}}{\text{R}-\text{CH}}-\text{CHO} \rightleftarrows \text{R}-\overset{\substack{\text{O}\\ \|}}{\text{C}}-\underset{\substack{|\\ \text{OH}}}{\text{CH}_2}$$

The mechanism of reaction (1) is unknown; it is speculated that it may involve the very slow ionization of the hydrated form of formaldehyde.

$$\text{HCHO} + \text{H}_2\text{O} \rightleftarrows \text{H}_2\text{C(OH)}_2 \rightarrow \text{HC}^-\text{(OH)}_2$$

$$\downarrow \text{HCHO} \mid \text{H}^+$$

$$\underset{\substack{|\\ \text{OH}}}{\text{CH}_2}-\text{CHO} + \text{H}_2\text{O} \rightleftarrows \underset{\substack{|\\ \text{OH}}}{\text{CH}_2}-\text{CH(OH)}_2$$

While the suggested synthesis of sugars from formaldehyde is attractive as a prebiotic process, it is not without difficulties.

1. Sugars are unstable in aqueous solution, particularly if the pH is much above 7. Under the conditions of the formose reaction they are destroyed in a time not very much longer than the time needed for their synthesis.
2. The mixtures of sugars obtained in the formose reaction usually contain only a small proportion of ribose.
3. So far, the reaction has not been demonstrated in solutions of formaldehyde less concentrated than 0.01 M. This may be due to inadequacies of the analytical techniques available; otherwise, the source of such concentrated formaldehyde solutions on the primitive earth could present a problem.

NUCLEOSIDE SYNTHESIS

Nucleoside synthesis under plausibly prebiotic conditions has proved to be unexpectedly difficult, so much so that no really satisfactory method has been reported. The difficulty is due in part to the existence of many isomers of each nucleoside, and in part to the poor nucleophilic reactivity of the bases relative to OH^-.

Ribose exists in two different ring forms, furanose (9a and 9b), and pyranose (10).

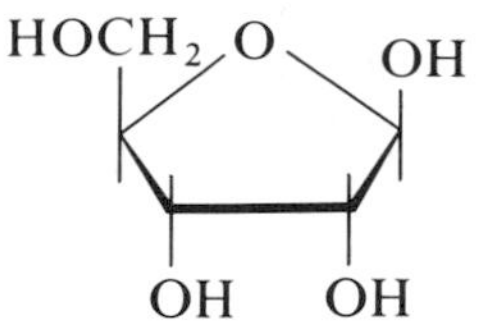

β-D-Ribofuranose
9a

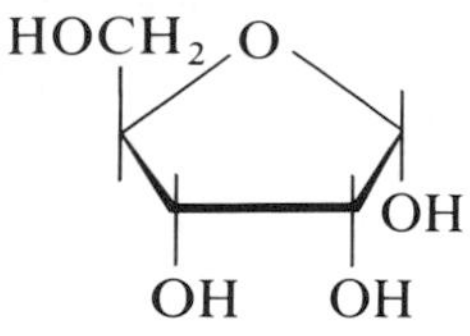

α-D-Ribofuranose
9b

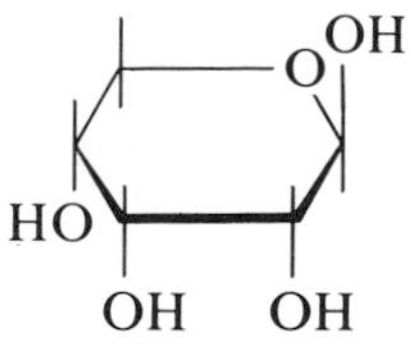

β-D-Ribopyranose
10

The natural nucleosides are all β-furanosides, but most organic syntheses give a mixture of isomers unless special precautions are taken. The heterocyclic bases can form bonds to a sugar at more than one position, five, for example, in the case of adenine. This leads to the formation of a multiplicity of products. It is necessary, therefore, to be particularly careful in the identification of natural nucleosides amongst the complex mixture of products formed in typical prebiotic reactions.

Direct heating of ribose with bases, either in the dry state or in aqueous solution, leads to little or no nucleoside synthesis. However, when purines are heated with ribose in the presence of certain inorganic salts, nucleosides are obtained in reasonable yield. The mixture of salts obtained by evaporating sea water is one of the most effective catalysts tested so far. With the help of this salt mixture, it has been possible to obtain 8% of β inosine and 2.3% of β adenosine. When a mixture of purine bases was heated with ribose and the salts from sea water, 4.5% β-inosine, 2% β-adenosine, 3.2% β-guanosine and 1.8% β-xanthosine were obtained. Uracil does not react under these conditions. Ribose reacts with the amino group of cytosine but does not form nucleosides. Ribose, adenine, and ethyl polyphosphate (or phenylpolyphosphate) give mixtures of α and β adenosine in good yields on heating as a semisolid mass, but these conditions cannot be considered prebiotic.

Attempts to obtain nucleosides by displacing other groups from the 1-position of a sugar have not led to plausible prebiotic syntheses. The classical syntheses of nucleosides using sugar halides in inert solvents do give good yields, but neither the reactants nor the solvents are plausibly prebiotic.

Two alternative approaches to the synthesis of nucleoside involve the building of a sugar on a preformed base, or, more plausibly, the construction of a base on a preformed sugar. The latter route has been used extensively in synthetic organic chemistry and has led to one model for prebiotic synthesis.

If ribose-5-phosphate is equilibrated with ammonia to give the ribosylamine, and then treated successively with cyanogen and cyanoacetylene, a material very similar to cytidylic acid is obtained in up to 20% yield. It is, in fact, almost entirely α-cytidylic acid rather than the natural β isomer. The same material is obtained in excellent yield from ribose-5-phosphate, cyanamide, and cyanoacetylene. The mechanism in this case involves the formation of an anhydronucleotide.

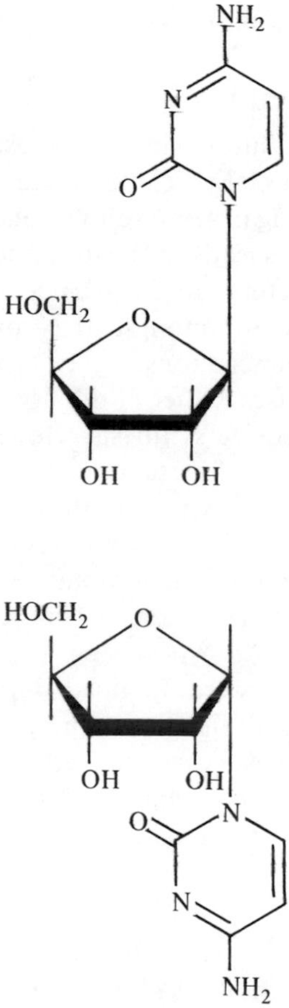

β-D-cytidine

α-D-cytidine

Fig. 8-2. The α and β isomers of cytidine.

The synthesis described above gives only very small amounts of the β-nucleotide, but this isomer can be obtained from the α isomer photochemically. It is well known that ultraviolet irradiation of β-cytidylic acid leads to the formation of a variety of products, including cytosine, a hydrate,

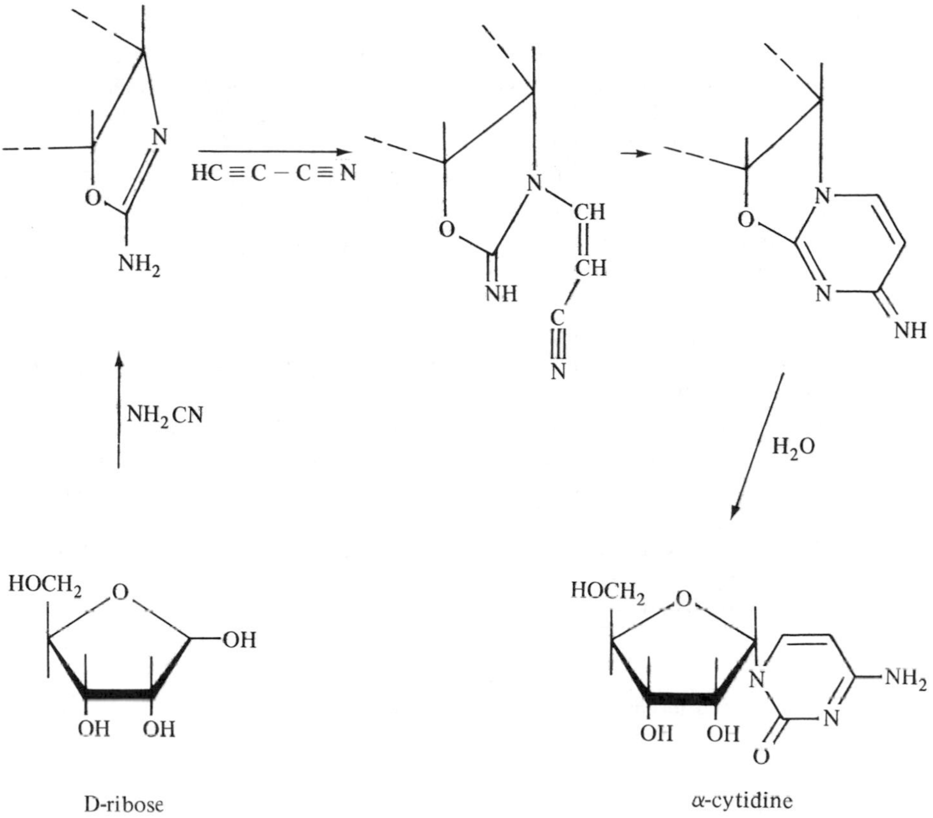

HOCH$_2$ D-ribose $\xrightarrow{NH_2CN}$ (oxazoline, NH_2) $\xrightarrow{HC \equiv C - C \equiv N}$ (N–CH=CH–C≡N, NH) $\rightarrow$ (anhydro, N, NH) $\xrightarrow{H_2O}$ α-cytidine

D-ribose

α-cytidine

Fig. 8-3. α-Cytidine synthesis. (Modified from Sanchez and Orgel, *J. Mol. Biol.* **47**, 542 (1970)).

and various uridylic acid derivatives. The irradiation of α-cytidylic acid gives as much as 10% of the natural β-cytidylic acid in addition to the products analogous to these from the ultraviolet irradiation of β-cytidylic acid.

The origin of nucleosides and nucleotides remains, in our opinion, one of the major problems in prebiotic synthesis. The proposed synthesis of free bases are reasonably plausible, and sugars can be obtained from formaldehyde in good yield under mild conditions. The instability of sugars in alkaline solution, and the difficulty of causing them to combine with the bases to give nucleosides is, however, a difficulty which remains to be overcome. Only the purine nucleosides have been obtained from free bases and ribose. It may be possible to produce nucleotides directly from sugars and simple precursors. However, no synthesis of adenosine along these lines has been achieved as yet, and the available synthesis of cytidylic acid is not completely convincing as a prebiotic process.

References

The synthesis of adenine from concentrated NH_4CN solutions was first shown by Oró and Kimball in 1960: J. Oró and A. P. Kimball, "Synthesis of adenine from ammonium cyanide," *Biochim. Biophys. Res. Commun.* **2**, 407 (1960).

J. Oró, "The mechanism of synthesis of adenine from hydrogen cyanide under possible primitive earth conditions," *Nature* **191**, 1193 (1961).

J. Oró and A. P. Kimball, "Synthesis of purines under possible primitive earth conditions. I. Adenine from hydrogen cyanide," *Arch. Biochem. Biophys.* **94**, 217 (1961).

J. Oró and A. P. Kimball, "Synthesis of purines under possible primitive earth conditions. II. Purine intermediates from hydrogen cyanide," *Arch. Biochem. Biophys.* **96**, 293 (1962).

This synthesis was confirmed by C. U. Lowe, M. W. Rees, and R. Markham, "Synthesis of complex organic compounds from simple precursors: formation of amino acids, amino acid polymers, fatty acids and purines from ammonium cyanide," *Nature* **199**, 219 (1963).

The synthesis of adenine by high-energy electrons was reported by C. Ponnamperuma, R. M. Lemmon, R. Mariner, and M. Calvin, "Formation of adenine by electron irradiation of methane, ammonia, and water," *Proc. Nat. Acad. Sci. U.S.* **49**, 737 (1963).

The kinetics and mechanisms of HCN polymerizations, as well as the properties of HCN trimer and tetramer are described in

R. A. Sanchez, J. Ferris, and L. E. Orgel, "Conditions for purine synthesis: Did prebiotic synthesis occur at low temperatures?" *Science* **153**, 72 (1966).

J. P. Ferris and L. E. Orgel, "Aminomalononitrile and 4-amino-5-cyanoimidazole in hydrogen cyanide polymerization and adenine synthesis," *J. Amer. Chem. Soc.* **87**, 4976 (1965).

J. P. Ferris and L. E. Orgel, "An unusual photochemical rearrangement in the synthesis of adenine from hydrogen cyanide," *J. Amer. Chem. Soc.* **88**, 1074 (1966).

J. P. Ferris and L. E. Orgel, "Aminomalononitrile and 4-amino-5-cyanoimidazole," *J. Amer. Chem. Soc.* **88**, 3829 (1966).

R. A. Sanchez, J. P. Ferris, and L. E. Orgel, "Synthesis of purine precursors and amino acids from aqueous hydrogen cyanide," *J. Mol. Biol.* **30**, 223 (1967).

R. A. Sanchez, J. P. Ferris, and L. E. Orgel, "Conversion of 4-aminoimidazole-5-carbonitrile derivatives to purines," *J. Mol. Biol.* **38**, 121 (1968).

Synthesis of pyrimidines from cyanoacetylene

R. A. Sanchez, J. P. Ferris, and L. E. Orgel, "Cyanoacetylene in prebiotic synthesis," *Science* **154**, 784 (1966).

J. P. Ferris, R. A. Sanchez, and L. E. Orgel, "Synthesis of pyrimidines from cyanoacetylene and cyanate," *J. Mol. Biol.* **33**, 693 (1968).

Sugars from formaldehyde

A. Butlerow, *Liebig's Ann. Chem.* **120**, 295 (1861); *Compt. Rend. Acad. Sci.* **53**, 145 (1861).

R. Mayer, K. Runge, and H. Dreschsel, *Z. für Chemie* **3**, 134 (1963).

N. W. Gabel and C. Ponnamperuma, *Nature* **216**, 453 (1967).

C. Reid and L. E. Orgel, *Nature* **216**, 455 (1967).

Nucleosides

R. A. Sanchez and L. E. Orgel, "Synthesis and photoanomerization of pyrimidine nucleosides," *J. Mol. Biol.* **47**, 531 (1970).

W. D. Fuller, R. A. Sanchez, and L. E. Orgel, "Synthesis of purine nucleosides," *J. Mol. Biol.* **67**, 25 (1972).

W. D. Fuller, R. A. Sanchez, and L. E. Orgel, "Solid-state synthesis of purine nucleosides," *J. Molec. Evolution* **1**, 249 (1972).

Chapter Nine

Stability of Organic Compounds and the Temperature of the Primitive Ocean

It is one thing to show that an organic compound can be synthesized under primitive-earth conditions, and it is quite another to be sure that it is sufficiently stable to have remained on the primitive earth long enough to have been available to the first organism. We have seen that life probably took one million to 1,000 million years to develop. Thus, either the compounds used were formed in a single event and had half-lives in excess of a million years or they were synthesized continuously. Of course, not all contemporary biochemicals had to be present in the first organism, so this principle applies only to essential compounds.

If compounds were synthesized continuously we must still explain how a sufficient concentration of critical intermediates built up. The steady state concentration would have depended on a balance between the rate of synthesis and the rate of decomposition. The rate of synthesis is limited by the amount of energy available so that only organic compounds with relatively long lifetimes could have accumulated in sufficient quantities to have been used in prebiotic syntheses. This argument is somewhat weakened when it is remembered that specific concentration mechanisms may have been operating for a number of prebiotic molecules; nonetheless, the problem of stability is an extremely important one.

It is worth pointing out that modern organisms have a very effective way of overcoming the problem posed by the instability of intermediates. Enzymes are present which increase by many orders of magnitude the rates of useful reactions while leaving the rates of decomposition reactions unchanged. Many of the more reactive intermediates of biochemistry would not have been useful in the primitive soup, since they would have undergone indiscriminate reaction with water and other substances in their environment.

Organic compounds are unstable in the presence of molecular oxygen. For example, butter turns bad (rancid) due to the oxidation of the carbon-carbon double bonds by molecular oxygen. Even at low levels of O_2, there is a slow oxidation of most organic compounds, and the rate is greatly enhanced in the presence of ultraviolet light. These and related arguments are so compelling that it does not seem possible that organic compounds remained in the primitive ocean for any length of time after O_2 entered the earth's atmosphere. They are now present on the surface of the earth only because they are continuously being resynthesized by living organisms. Organic compounds occur below the surface of the earth, for example in coal and oil, because there the environment is anaerobic.

It appears certain, then, that O_2 was absent from the earth's atmosphere during the period when organic compounds were synthesized and probably up to the time when the first organism evolved. Therefore, we will summarize what is known about the stabilities of biologically interesting organic compounds under anaerobic conditions. These data will allow us to draw some conclusions about the compounds which could have been present in the first organism, as well as to make some comments on the temperature of the primitive ocean.

AMINO ACIDS

The 20 amino acids that occur in proteins have greatly different stabilities. The most stable are the aliphatic amino acids, glycine, alanine, valine, isoleucine, and leucine. They decompose by irreversible decarboxylation; for example,

$$CH_3{-}\underset{\displaystyle NH_3^+}{\underset{|}{CH}}{-}COO^- \rightarrow CH_3CH_2{-}NH_2 + CO_2$$

The rate has been investigated in some detail for alanine. The half-life is about 20×10^9 years at 0°C, 3×10^9 years at 25°C, and 10 years at 150°C. The other aliphatic amino acids are also quite stable on the geological time scale.

Proline and lysine are predicted to have a stability to decarboxylation comparable to that of the aliphatic amino acids, but no quantitative data are available. Phenylalanine decarboxylates with a half-life of 10^9 years at 0°C, 10^8 years at 25°C, and 10^4 years at 100°C, but tyrosine decomposes much faster. Histidine is relatively unstable, particularly to acid hydrolysis, but little quantitative information is available.

The remaining amino acids all have alternative mechanisms of decomposition in addition to decarboxylation. Arginine undergoes hydrolysis at a moderate rate in neutral solution, either directly to ornithine or first to citrulline and then to ornithine.

$$H_2N-\overset{\overset{\displaystyle NH}{\|}}{C}-NH-CH_2-CH_2-CH_2-\underset{\underset{\displaystyle NH_2}{|}}{CH}-COOH \rightarrow$$

$$H_2N-\overset{\overset{\displaystyle O}{\|}}{C}-NH-CH_2-CH_2CH_2-\underset{\underset{\displaystyle NH_2}{|}}{CH}-COOH$$

$$\rightarrow H_2N-CH_2-CH_2CH_2-\underset{\underset{\displaystyle NH_2}{|}}{CH}-COOH$$

We estimate that the half-life would be a few thousand years at 25°C. Glutamic acid first cyclizes to pyrrolidone-α-carboxylic acid (pyroglutamic acid), which subsequently decarboxylates slowly.

$$^{-}OOC-CH_2-CH_2-\underset{\underset{\displaystyle NH_3^+}{|}}{CH}-COO^- \rightarrow$$

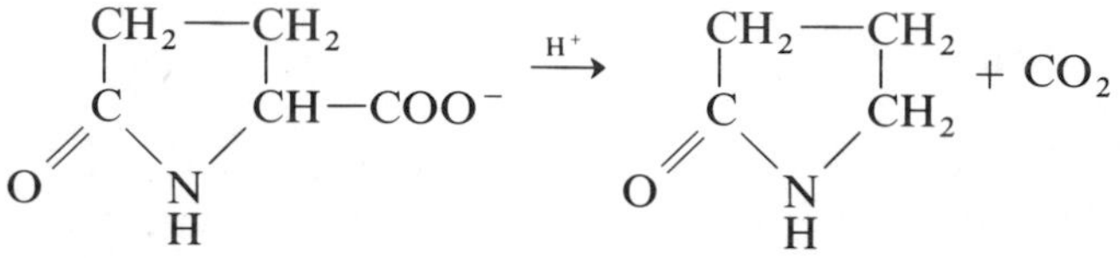

$$\text{(ring: } CH_2-CH_2,\ C(=O),\ CH-COO^-,\ NH\text{)} \xrightarrow{H^+} \text{(ring: } CH_2-CH_2,\ C(=O),\ CH_2,\ NH\text{)} + CO_2$$

Aspartic acid decarboxylates at high temperatures to give a mixture of α- and β-alanine, but at low temperatures the predominant reaction is deamination.

$$^{-}OOC-CH_2-\underset{\underset{\displaystyle NH_3^+}{|}}{CH}-COO^- \rightleftarrows {}^{-}OOC-CH=CH-COO^- + NH_4^+$$

This decomposition is largely reversed by 0.01 M NH_4^+. In Chapter 4 we attempted to use equilibrium data for the deamination of aspartic acid to estimate the ammonia concentration in the primitive ocean.

Serine and threonine have short lives with respect to decarboxylation: 10^6 years at 0°C, 10^3 years at 25°C, and 1 year at 100°C. In addition, they decompose by the following alternate routes, which are probably more important.

$$R\underset{\underset{\displaystyle OH}{|}}{CH}-\underset{\underset{\displaystyle NH_2}{|}}{CH}-COOH \rightarrow RCHO + \underset{\underset{\displaystyle NH_2}{|}}{CH_2}-COOH \qquad \begin{array}{l} R = H \text{ Serine} \\ R = CH_3 \text{ Threonine} \end{array}$$

$$\underset{\text{OH}}{\text{RCH}}-\underset{\text{NH}_2}{\text{CH}}-\text{COOH} \xrightarrow{-\text{H}_2\text{O}} \text{R}-\text{CH}=\underset{\text{NH}_2}{\text{C}}-\text{COOH}$$

$$\xrightarrow[-\text{NH}_3]{+\text{H}_2\text{O}} \text{R}-\text{CH}_2-\overset{\text{O}}{\overset{\|}{\text{C}}}-\text{COOH}$$

The first pathway is reversible and provides a potential prebiotic synthesis. Unless a mechanism for the rapid synthesis of these amino acids in large amounts can be proposed, their rapid decomposition at elevated temperatures is a strong argument against a hot environment for the origins of life.

The instability of cysteine and methionine is mainly due to reactions with oxygen. However, even under anaerobic conditions they are not very stable. Cysteine decomposes with the elimination of hydrogen sulphide.

$$\text{Cysteine} \rightarrow \text{pyruvate} + \text{NH}_3 + \text{H}_2\text{S}$$
$$\searrow \text{glycine} + \text{H}_2\text{CO} + \text{H}_2\text{S}$$

Since the reaction to give glycine may be reversible, this also suggests a possible prebiotic synthesis. Methionine is unstable in the presence of oxygen, but is relatively stable under anaerobic conditions. However, some prebiotic reagents (e.g., cyanogen and cyanogen chloride) might destroy the methionine.

Glutamine decomposes so rapidly to give pyroglutamic acid that it is doubtful that any significant concentration could have occurred in the primitive ocean. On the other hand, glutamine in a peptide linkage is relatively stable. If glutamine occurred in primitive proteins, it must have been incorporated by a very rapid synthesis using a small concentration of available glutamine; it could also have been made from glutamic acid already incorporated in a peptide.

Asparagine is unstable with respect to hydrolysis to aspartic acid and (like aspartic acid) with respect to deamination.

$$\text{H}_2\text{N}-\overset{\text{O}}{\overset{\|}{\text{C}}}-\text{CH}_2-\underset{\text{NH}_3^+}{\text{CH}}-\text{COO}^- \nearrow {}^-\text{OOC}-\text{CH}_2-\underset{\text{NH}_3^+}{\text{CH}}-\text{COO}^- + \text{NH}_4^+$$
$$\searrow \text{H}_2\text{N}-\overset{\text{O}}{\overset{\|}{\text{C}}}-\text{CH}=\text{CH}-\text{COO}^- + \text{NH}_4^+$$

The free amide group of asparagine in polypeptides also seems to be relatively unstable with respect to hydrolysis.

Tryptophan is unstable in acid and rarely survives acid hydrolysis of protein. It is stable to base hydrolysis. In the range most interesting for prebiotic chemistry, tryptophan appears to be reasonably stable above pH 8, but unstable below pH 6. In the presence of formaldehyde, several unstable amino acids are converted to stable derivatives. Tryptophan reacts to form 3,4,5,6-tetrahydrocarboline-5-carboxylic acid.

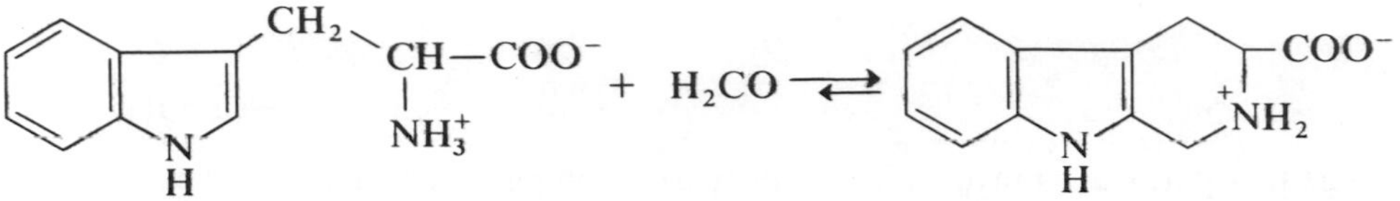

$$\text{Tryptophan} \; (CH_2\text{–}CH(NH_3^+)\text{–}COO^-) + H_2CO \rightleftarrows \text{tetrahydrocarboline} \; (COO^-, \overset{+}{N}H_2)$$

The tetrahydrocarboline appears to be more stable than tryptophan itself. Since this reaction is reversible, it is possible that the tryptophan was converted to the tetrahydrocarboline and was decomposed back to tryptophan when free formaldehyde was no longer abundant in the primitive ocean. This tryptophan would then have been available for prebiological peptide synthesis, or for use by early organisms. This example has been presented to show that molecules that are unstable in water may have been stabilized in the prebiotic environment.

FATTY ACIDS

Fatty acids are among the most stable organic compounds, especially in the absence of O_2. They are more stable than amino acids with respect to decarboxylation. Therefore, if conditions were such that amino acids were stable, the fatty acids would have been stable also.

SUGARS

Sugars, like other hydroxyaldehydes, are unstable in basic solution. The hexoses, for example, decompose in 1 M NaOH through a series of intermediate compounds to give lactic acid as the major product. Sugars also react relatively rapidly in neutral solutions with various other inorganic and organic compounds. In particular, they react with amino acids to form several brown products. This is the basis of the browning (or Maillard) reaction, which, for example, causes apples to turn brown on exposure to air.

The formation of the brown colored products is accelerated by O_2, but there is still a reaction between the aldehyde of the sugar and the amino acid under anaerobic conditions. Whether this reaction is reversible or not is unknown. It would be very useful to have experimental data, measured under anaerobic conditions, on the rate of decomposition of sugars at pH values around neutrality. Similar data on the reaction of sugars with various compounds such as amino acids would also be useful.

It is possible that sugars did not occur in the primitive ocean as free compounds but rather in a combined form. Although sugars are very important in present day energy metabolism and are present in cell walls and other structural elements, their only essential genetic role is in the nucleic acids. Therefore, the first organisms may have used sugars only for their DNA and/or RNA.

When ribose is combined in the form of nucleotides its reactivity is greatly diminished. The stability of the sugars in nucleic acids is not the limiting factor for the stability of the polymers. Thus, if we could establish that nucleosides occurred in the primitive ocean we could neglect any problems arising from the instability of the free sugars. However, at present, no completely satisfactory prebiotic synthesis of nucleosides has been described.

PURINES, PYRIMIDINES, AND NUCLEOSIDES

The rates of hydrolysis of adenine to hypoxanthine and of guanine to xanthine seem to be so slow at reasonable pH values that the instability of these purines is not a serious problem. The pyrimidine ring of thymine, uracil, and cytosine is also quite stable. However, cytosine hydrolyzes to uracil rather rapidly, and cytidine is hydrolyzed to uridine at a similar rate (a half-life less than 20,000 years under reasonable prebiotic conditions). This is a real difficulty if it is assumed that cytosine was required for nucleic acids in the first organism. Perhaps cytidine or cytosine was resynthesized from uridine or uracil. This does not seem to be a particularly difficult reaction, but no one has demonstrated it. Alternatively, cytosine may have been protected from hydrolysis in the form of a double-stranded polynucleotide. Finally, of course, the first nucleic acids may not have contained cytosine; they may, for example, have contained only adenosine and uridine.

The nucleosides and deoxynucleosides are unstable with respect to hydrolysis to the base and ribose or deoxyribose. The rate of this reaction is slow for the pyrimidine nucleosides. Uridine and 5-methyl uridine hydrolyze at a constant rate between pH 5 and 7; OH^- has a weak catalytic effect between pH 7 and 10. Deoxyuridine and thymidine hydrolyze at a rate independent of pH between pH 5 and 9. The kinetic data are not good enough to permit accurate extrapolations to lower temperatures from the measurements that were made at temperatures between 75 and 135°C, but very approximate half-lives are:

Compound	Half-Life at 0°C (yr)	Half-Life at 25°C (yr)
Uridine	1.6×10^6	1.2×10^5
5-Methyl uridine	1.6×10^6	1.2×10^5
Deoxyuridine	6×10^4	365
Thymidine	2×10^5	10^3

As we have seen, cytidine and deoxycytidine hydrolyze first to uridine and deoxyuridine, with a half-life of less than 10^3 years under prebiotic conditions.

The purine ribonucleosides are less stable than the pyrimidine nucleosides to hydrolysis in acidic or basic solution. The rate in neutral solutions is not known, but it is probably greater than for the pyrimidine nucleosides. The purine deoxyribosides are very unstable to acid hydrolysis; even at pH 8.5, *acid* hydrolysis is fast enough to destroy them in the times with which we are concerned.

THE HYDROLYSIS OF POLYPEPTIDES AND POLYNUCLEOTIDES

The first step in the determination of the composition of a protein is usually acid hydrolysis, although sometimes alkaline hydrolysis is used. Few experiments measuring the rate of hydrolysis in the pH range of 5 to 10 seem

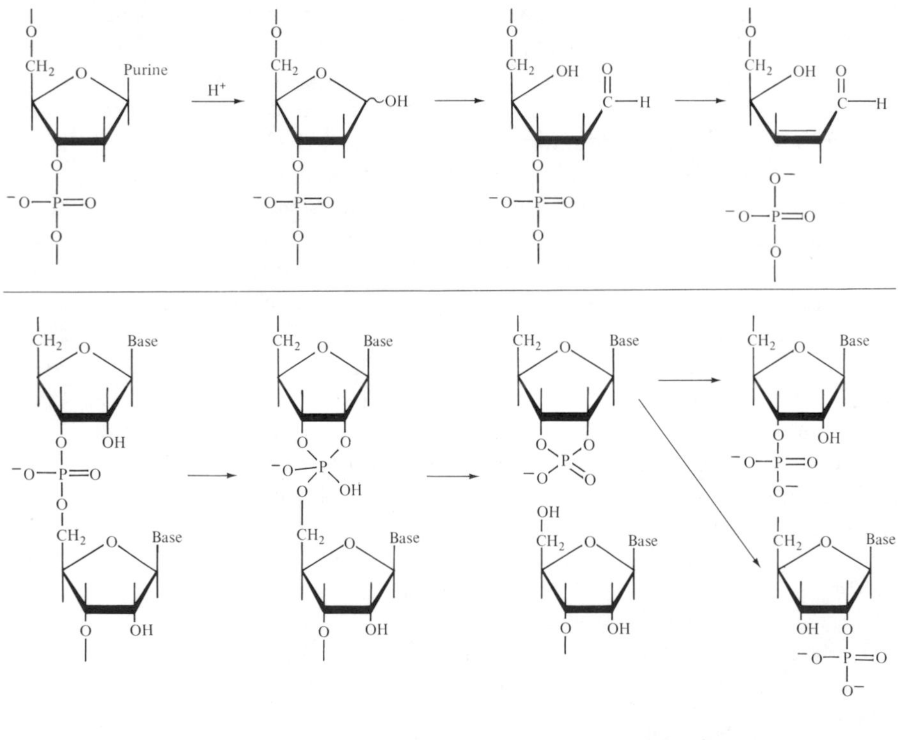

Fig. 9-1. The mechanism of hydrolysis DNA by a β-elimination reaction (upper) and the mechanism of hydrolysis of RNA by formation of a 2′,3′-cyclic phosphate (lower).

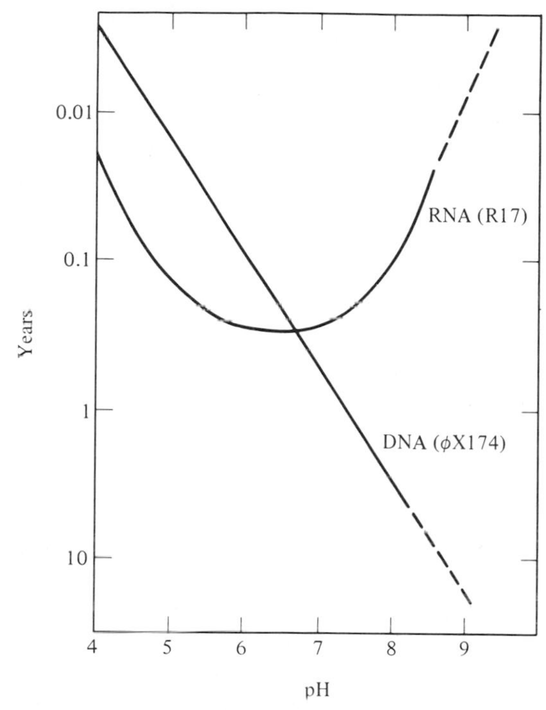

Fig. 9-2. Half-lives for the depurination of the single-stranded DNA virus ϕX174 and half-lives for the hydrolysis of phosphodiester bonds of the single-stranded RNA virus R17 at 70°C. The data are based on rates of inactivation of virus infectivity (Ginosa, *et al.*, *Nature* **203**, 606, 1964). The dotted lines are extrapolations of the experimental data. At higher pH values, the dotted line for DNA would bend upwards due to hydrolysis of the phosphodiester bond.

to have been reported. The kinetic data on alkaline hydrolysis are quite limited, but it is possible to extrapolate from results in alanylalanine at pH 14 and high temperatures to pH 8 and moderate temperatures. The half-lives are 8×10^7 years at 0° and 6×10^5 years at 25°C. The extrapolation over such a wide range of pH is not very reliable, so the half-lives may be shorter than this, particularly if a change in the hydrolysis mechanism occurs between pH 14 and pH 8.

Deoxyribonucleic acids are very sensitive to acid hydrolysis; ribonucleic acids to basic hydrolysis. Both are hydrolyzed in strong acid or strong base. DNA hydrolyzes by initial depurination. The depurinated deoxyribose then breaks the chain by a β-elimination reaction. RNA hydrolyzes by breaking the phosphodiester linkage. The reaction involves a 2′–3′ cyclic phosphate as an intermediate.

The data in Fig. 9-2 are based on the rate of inactivation at 70°C of a single-stranded DNA virus, ϕX174, and a single-stranded RNA virus, R17. The inactivation appears to be due to depurination for the DNA and breaking of phosphodiester bonds for the RNA. The rates have been recalculated to show average half-lives for *each* purine deoxynucleoside residue and *each* phosphodiester bond. Extrapolation of these data gives half-lives at pH 7.15 of

Residue	Half-Life at 0°C (yr)	Half-Life at 25°C (yr)
Depurination of DNA	23,000	360
Hydrolysis of phosphodiester bonds of RNA	900	30

The pH of maximum stability of RNA is about 6.5. It is interesting that DNA has a similar rate of decomposition at this pH.

The data discussed above refer to single-stranded molecules in the absence of calcium and magnesium. Divalent cations catalyze the hydrolysis of RNA so that half-lives on the primitive earth may have been much lower. Double-stranded DNA is more stable to depurination than single-stranded DNA by at least a factor of 10, and this may have been important for the accumulation of DNA. These complications make it impossible to estimate relative half-lives under primitive-earth conditions.

We have been discussing the stability of organic molecules in aqueous solution. It must be realized, however, that some compounds that hydrolyze rapidly in solution are quite stable as dry solids. It is possible that some of the compounds which are too unstable to have accumulated in aqueous solution were deposited as solids and then survived for long periods. Unfortunately, very little is known about the long term stability of organic solids.

THE TEMPERATURE OF THE PRIMITIVE OCEAN

We have seen that the half-lives of a number of compounds of biological importance are less than a million years at 25°C. Since the temperature coefficients of these decompositions are large, the half-lives would be much less at temperatures of 50° or 100°C. Conversely, the half-lives would be longer at 0°C. The rates of depurination of DNA, of hydrolysis of peptide and polynucleotide polymers, and of decomposition of sugars, are so large that it seems impossible that such compounds could have accumulated in aqueous solution and have been used in the first organism, unless the temperature was low.

The temperature of the present ocean averages 4°C, with the surface waters somewhat warmer. While the freezing point of water is 0°C, the freezing point of sea water is −1.8°C. Sea water solidifies almost completely at −21°C. We do not know what the temperature was in the primitive ocean, but we can say that the instability of various organic compounds and polymers makes a compelling argument that life could not have arisen in the ocean unless the temperature was below 25°C. A temperature of 0°C would have helped greatly, and −21°C would have been even better. At such low temperatures, most of the water on the primitive earth would have been in the form of ice, with liquid sea water confined to the equatorial oceans.

There is another reason for believing that life evolved at low temperatures, whether in the oceans or lakes. All of the template-directed reactions that must have led to the emergence of biological organization take place only below the melting temperature of the appropriate organized polynucleotide structure. These temperatures range from 0°C, or lower, to perhaps 35°C, in the case of polynucleotide-mononucleotide helices (see Chapter 11).

The environment in which life arose is frequently referred to as a warm, dilute soup of organic compounds. We believe that a cold, concentrated soup would have provided a better environment for the origins of life. At low temperatures the decomposition of organic compounds and polymers is slowed down greatly. Furthermore, at low temperatures the concentration of organic compounds of the eutectic-freezing mechanism could have occurred. Although at first sight low temperatures might seem to have been a disadvantage (chemical syntheses would have proceeded more slowly) in fact they may have been advantageous. It is the ratios of the rates of synthesis to the rates of decomposition which are important, rather than the absolute rates, if ample time was available. Since the temperature coefficients of the synthetic reactions are generally less than those for the decomposition reactions, low temperatures would have favored the synthesis of more complex organic compounds and polymers.

References

J. R. Vallentyne, *Geochim. Cosmochim. Acta* **28**, 157 (1964). Decarboxylation of alanine, serine, threonine, phenylalanine, and pyroglutamic acid.

D. Povoledo and J. R. Vallentyne, *Geochim. Cosmochim. Acta* **28**, 731 (1964). Decarboxylation of pyroglutamic acid.

J. R. Vallentyne, *Geochim. Cosmochim. Acta* **32**, 1353 (1968). Decarboxylation of proline and leucine, and decomposition of lysine and arginine.

Murray *et al.*, *J. Biol. Chem.* **240**, 705 (1965). Decomposition of arginine to citrulline and ornithine.

J. L. Bada and S. L. Miller, *J. Am. Chem. Soc.* **92**, 2774 (1970). Deamination of aspartic acid and asparagine.

N. Friedmann, W. J. Haverland, and S. L. Miller, in *Chemical Evolution and the Origin of Life*, R. Buvet and C. Ponnamperuma, eds. (North Holland, Amsterdam, 1971), p. 123. Synthesis of serine from formaldehyde and glycine and its reversal.

H. Zahn and E. Golsch, *Z. Physiol. Chem.* **330**, 38 (1962); C. De Marco, M. Coletta, and D. Cavallini, *Arch. Biochem. Biophys.* **100**, 51 (1963); O. Gawron and G. Odstrchel, *J. Am. Chem. Soc.* **89**, 3263 (1967). Decomposition of cysteine and related compounds.

R. B. Martin, A. Parcell, and R. I. Hedrick, *J. Am. Chem. Soc.* **86**, 2406 (1964); P. B. Hamilton, *J. Biol. Chem.* **158**, 375 (1945). Cyclization of glutamine to pyroglutamic acid.

R. Shapiro and R. S. Klein, *Biochemistry* **5**, 2358 (1966); **6**, 3576 (1967); J. P. Ferris, R. A. Sanchez, and L. E. Orgel, *J. Mol. Biol.* **33**, 693 (1968). Deamination of cytosine and cytidine to uracil and uridine.

R. Shapiro and S. Kang, *Biochemistry* **8**, 1806 (1969). Hydrolysis of deoxyuridine and thymidine. S. L. Miller (unpublished data). Hydrolysis of uridine and 5-methyl uridine.

H. Hartmann *et al.*, *Z. Naturforschung* **17b**, 143 (1962). P. A. Levene *et al.*, *J. Biol. Chem.* **82**, 167 (1929). Rates of hydrolysis of peptides in base.

W. Ginoza *et al.*, *Nature* **203**, 606 (1964). J. Eigner, H. Boedtker, and G. Michaels, *Biochim. Biophys. Acta* **51**, 165 (1961). Hydrolysis of DNA and RNA at pH's between 5 and 10. D. A. Usher, D. I. Richardson, Jr., and D. G. Oakenfull, *J. Am. Chem. Soc.* **92**, 4699 (1970). Hydrolysis of a ribonucleoside analog.

Chapter Ten

Concentration Mechanisms

We have mentioned that the primitive ocean is often described as a warm, dilute soup, but in the last chapter we showed that a cold ocean would have provided a more advantageous environment for the accumulation and use of prebiotic organic compounds. Nor is it obvious that the soup was very dilute. If all the carbon on the surface of the earth (3,000 g cm^{-2}) were dissolved as organic compounds in the present oceans (274 kg cm^{-2}), the resulting solution would contain 1% organic carbon.

It is very unlikely that all the surface carbon was ever present as organic compounds, but 1 to 20% might have been. If the primitive oceans were as large as the present oceans, this would have given a solution 0.003 to 0.03 M in typical prebiotic organic compounds (C_3 compounds, for example). If the primitive oceans were smaller, the concentration would of course have been higher. These relatively concentrated solutions would not have been in chemical equilibrium. They would have persisted in a steady state only because spontaneous decomposition was balanced by resynthesis in the atmosphere and oceans.

Some prebiotic syntheses would certainly have occurred in the primitive oceans, but others would have been restricted to regions where the concentrations were higher. The efficient synthesis of polymers, in particular, is not likely to have taken place in the bulk of the oceans. Thus, it is important to consider concentration mechanisms which might have led to the building up of more concentrated solutions in localized regions on the earth's surface.

The simplest mechanism that might have operated on the primitive earth is evaporation. This would have been most important in lakes, since solutions of nonvolatile solids, no matter how dilute initially, can be concentrated until the solubility limit is reached, and the solid precipitates. The ratio of the final to the initial concentration (the concentration factor) can be very large. Precipitation itself can also be considered a concentration mechanism.

In tidepools, the situation is rather different. Evaporation concentrates salts as well as organic compounds, and sooner or later the salts begin to precipitate. If an organic compound is soluble in saturated salt solutions it can continue to concentrate, but a large amount becomes absorbed on, or trapped in, the mass of solid salt. Hence, organic compounds are not concentrated as efficiently in salt water as in fresh water.

Compounds such as hydrogen cyanide, formaldehyde, and acetaldehyde cannot be concentrated by evaporation because they are more volatile than water. Evaporation of a lake containing these compounds would lower their concentration rather than raise it. It is interesting that all of the most important structural components of living organisms such as amino acids, purines, sugars, and higher fatty acids are nonvolatile. By contrast, many of their prebiological precursors, such as cyanogen, cyanide, and aldehydes, are volatile. Ammonia is more volatile than water, but in neutral or acid solution it is present mainly as nonvolatile ammonium salts. The ammonium salts could have been concentrated by evaporation in many environments on the primitive earth.

Organic compounds can often be concentrated by adsorption on surfaces —for example, on the surfaces of clays and other minerals, or even at air-water interfaces. Adsorption on solids must have been important in prebiological syntheses, but as yet there are only a few good examples in which it has been shown to be significant. One example, which will be discussed in the next chapter, is the formation of peptides from aminoacyladenylates on Montmorillonite.

We think that the later stages of biochemical evolution are particularly likely to have occurred on surfaces, since polymers are, in general, much more strongly adsorbed than monomers. Surfaces would tend to "fractionate" polypeptide or polynucleotide solutions, adsorbing the longer oligomers and leaving monomers in solution.

A potentially prebiotic surface reaction of a somewhat different kind is the synthesis of pyrophosphate on hydroxylapatite [$Ca_{10}(PO_4)_6(OH)_2$] using cyanate as a condensing agent (Chapter 11). The maximum yield in the presence of apatite is as high as 26%, while 1/2% or less is formed in a 0.5 M solution of a soluble phosphate.

It is difficult to understand how the activation of phosphate could have occurred to a significant extent in solution. The concentration of phosphate in the oceans is low (3×10^{-6} M or less) because calcium phosphate (hydroxylapatite) is insoluble. Since apatite has probably always been the major phosphate mineral on the floor of the ocean, it is likely that many reactions of inorganic phosphate took place on the surface of hydroxylapatite.

The absorption of organic material in soluble colloidal particles, called *coacervates*, has been proposed as an important concentration mechanism for the primitive earth. Reactions in coacervates are closely related to reactions on surfaces, except that dissolved material can be absorbed throughout the body of the coacervate. Many interesting studies of coacer-

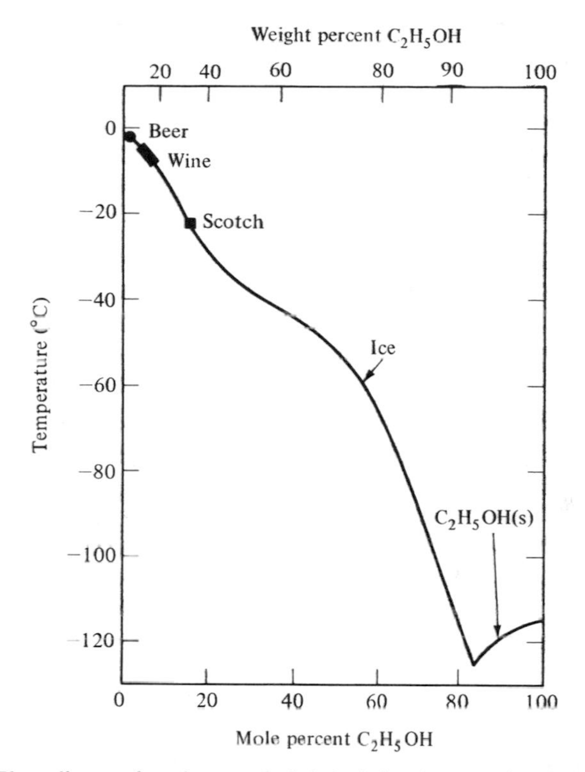

Fig. 10-1. Phase diagram for mixtures of ethyl alcohol and water. If a solution containing 40 M mole % C_2H_5OH is cooled, ice begins to separate at −42° as shown by the solid line marked "ice." If a solution containing 90% C_2H_5OH is cooled, solid C_2H_5OH begins to separate at approximately −120°C. Temperatures at which beer, wine, and Scotch are in equilibrium with ice are indicated. The eutectic is at −125°C and contains 83 mole % C_2H_5OH.

vates have been reported, particularly by Russian workers, but it is not yet possible to say how they were involved in prebiotic synthesis.

Another important concentration mechanism involves the freezing of aqueous solutions. As more and more ice separates, solutes which are unable to enter the crystal lattice of ice are left in solution. The production of applejack by the freezing of cider is a classical example of this process. Temperatures in many parts of the U.S. and Canada are such that a barrel of cider will "freeze" in a few days, except for a small liquid residue which contains a satisfactorily high proportion of ethyl alcohol.

Any compound can be concentrated in this way to an extent which is limited only by its solubility in water at low temperatures. Concentration factors have little meaning in this context, since the terminal concentration is independent of the initial concentration.

Consider a simple example, the freezing of a 1.0 molal (m) NaCl solution. Ice begins to separate at −3.55°C, and then as the temperature is lowered

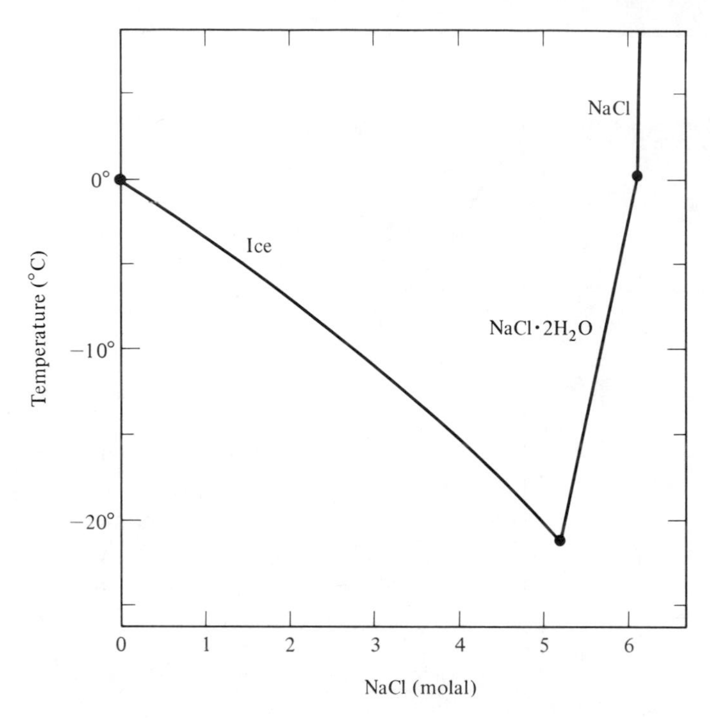

Fig. 10-2. Phase diagram for aqueous sodium chloride. Above 0.15°C, NaCl is the stable solid; the dihydrate is formed at lower temperatures. The eutectic temperature is −21.12°C and the eutectic concentration is 5.20 molal (23.31 wt. % NaCl).

the concentration of salt increases. At −7.20°C about half of the water has frozen out and a 2 m solution remains. The NaCl solution continues to concentrate until, at −21.1°C, the eutectic temperature is reached. At this temperature the whole mass solidifies to give a mixture of ice and salt. The sodium chloride is present as a solid hydrate, $NaCl \cdot 2H_2O$, but some substances separate in unhydrated form. At the eutectic temperature, the concentration of the salt solution is 5.2 m. This is known as the *eutectic concentration.* In this example, the concentration factor is 5.2, but we could equally well have made it 52 or 5,200 by starting with 0.1 M or 0.001 M salt solution. Our example emphasizes the point that it is the terminal concentration rather than the concentration factor which is most important in discussions of eutectic freezing.

In Chapter 7 we saw that the polymerization of cyanide to its tetramer is a step in the synthesis of purines and certain amino acids. This step is carried out advantageously by making use of the eutectic-concentration mechanism.* The eutectic temperature of hydrogen cyanide-water mixtures is −21°C and the eutectic contains 75 wt % HCN. Although synthesis is

* We will use the expression "eutectic concentration" to include both concentration at the eutectic temperature and concentration by freezing at higher temperatures.

slower below 0°C, it still proceeds at a rate of a few percent per month. Suppose we start with a 0.001 M solution of cyanide, adjusted to pH 8.5. At any temperature in the range 0–100°C, hydrolysis to formic acid accounts for 90–99% of the starting material. Cooling to −20°C yields a very concentrated solution in which hydrolysis is insignificant compared to polymerization, although the latter takes many months.

Many important biochemicals, including many amino acids, sugars, and nucleotides, are extremely soluble in water and can be concentrated extensively by freezing. So far, few prebiotic reactions have been studied in detail in eutectic solutions, but preliminary results on the synthesis of nucleotides and oligonucleotides have been reported. Many other syntheses may be possible in eutectic solutions.

The freezing of large volumes of salt water is more complicated than we have indicated above. When sea water freezes, a substantial amount of salt solution is trapped between the ice crystals, although no salt is incorporated in them. The water obtained by melting fresh sea ice is too salty for drinking. However, as sea ice ages, the entrapped salt solution slowly works its way down through the ice crystals leaving the top of the ice pack free of salt. Experienced residents in Arctic regions are aware of this and use the less freshly formed ice, which is sometimes called *rotten ice.* It is possible that, on the primitive earth, compounds trapped in the ice might have separated in a similar way.

If two bodies of water at different temperatures are in equilibrium with the same atmosphere, materials in the atmosphere will dissolve to a greater extent in the cooler body of water. Consider two aqueous hydrogen cyanide solutions in equilibrium with a rapidly mixed atmosphere. Since the partial pressure of hydrogen cyanide is the same everywhere, the concentration of HCN in the two solutions must adjust so that they have the same equilibrium vapor pressure. Suppose the total amount of HCN present is such that the partial pressure of HCN is 5×10^{-4} atm. In a solution at 0°C, the HCN concentration would have to be 1.1×10^{-2} M. Hydrogen cyanide is much more volatile at 35°C, so a solution at that temperature would contain only 2.9×10^{-3} M HCN. Thus, a concentration factor of 3.8 would be achieved in this case. Concentration gradients of this kind, which are not particularly large, must have existed on the primitive earth.

One final possibility is that concentration was effected with the help of membranes. Living cells can concentrate metabolites efficiently from very dilute solution and then retain them within their membranes. The free energy required to achieve this result is supplied by the hydrolysis of ATP. The coupling of ATP hydrolysis to the flow of metabolites through the membrane is achieved by a series of proteins called *permeases.* It seems unlikely that similar concentration mechanisms operated before life was well established, but this is not certain.

Although an inert membrane cannot, by itself, concentrate a solute, it can act as a barrier to prevent mixing of solutions. Phospholipids for example readily form selectively permeable membranes which act as barriers to the flow of Na^+, K^+, and certain other ions. The artificial lipid membranes which have been prepared so far are mechanically fragile, but more robust membranes could probably be made. Perhaps such membranes played some role in chemical evolution.

References

H. C. Urey, *Proc. Nat. Acad. Sci. U.S.* **38**, 351 (1952). The possible concentrations of organic compounds in the primitive oceans.

S. L. Miller and M. Parris, *Nature* **204**, 1248 (1964). Synthesis of pyrophosphate from cyanate and apatite.

R. A. Sanchez, J. P. Ferris, and L. E. Orgel, *Science* **153**, 72 (1966). Concentration of HCN by freezing and synthesis of HCN tetramer.

A. I. Oparin, *The Origin of Life* (Dover, New York, 1953). Discussions of coacervates. Other editions of Oparin's books also have extensive discussions of coacervates.

Chapter Eleven

Dehydration Reactions and Polymerizations

The most important macromolecules in living systems are all subject to hydrolysis in aqueous solution. The persistence of high concentrations of proteins and nucleic acids in living cells is the result of a favorable balance between the rate of their biosynthesis from activated intermediates and the rate of their hydrolysis to nonactivated components. Living systems are always far from being in equilibrium with their environment. The existence of proteins and nucleic acids is one manifestation of this disequilibrium.

In contemporary biochemistry, ATP is the most important energy-rich intermediate used in the synthesis of polymers. The pyrophosphate bonds are generated either at the expense of free energy released in chemical reactions, such as the oxidation of carbohydrates and other organic compounds, or more indirectly from sunlight in the process of photophosphorylation. The free energy of the pyrophosphate bonds of ATP is coupled to dehydration reactions by means of enzymes.

$$\begin{array}{ccc} \text{ATP} & & \text{Monomers} \\ \downarrow & \text{Enzymes} & \downarrow \\ \text{AMP} + P_2O_7^{4-} & & \text{Polymers} \end{array}$$

The two most important reactions are peptide bond formation

$$H_2N{-}\underset{\displaystyle R_1}{\overset{}{CH}}{-}COOH + H_2N{-}CH(R_2){-}COOH \rightarrow H_2N{-}CH(R_1){-}C(=O){-}NH{-}CH(R_2){-}COOH$$

and nucleotide bond formation

$$\begin{array}{c} \text{Base}_1 \\ | \\ \text{Ribose-5}'\text{-phosphate} \end{array} + \begin{array}{c} \text{Base}_2 \\ | \\ \text{Ribose-5}'\text{-phosphate} \end{array} \longrightarrow \begin{array}{c} \text{Base}_1 \qquad\qquad\qquad \text{Base}_2 \\ | \qquad\qquad\qquad\qquad | \\ \text{Ribose-5}'\text{-phosphate-3}'\text{-Ribose-5}'\text{-phosphate} \end{array}$$

The formation of nucleosides from bases and sugars, of nucleotides from nucleosides and phosphate, and of lipids and polysaccharides from their components, are all uphill reactions.

Many attempts have been made to carry out these same reactions under prebiotic conditions. The simplest of all dehydrating mechanisms is the elimination of the elements of water from a dry solid by direct heating. Such thermal methods have been used extensively in models of protein and nucleic acid synthesis. A modification which has been studied is the heating of monomers in a dehydrating solvent such as polyphosphoric acid. A totally different approach involves the coupling of the hydrolysis of an energy-rich compound with the dehydration reaction in aqueous solution. This comes much closer to the biochemical mechanisms of polymer synthesis. A third approach attempts the synthesis of energy-rich precursors which can polymerize spontaneously; for example, the nitriles or the amides of the amino acids can be polymerized to peptides. Finally, a few reports have appeared describing the use of visible or ultraviolet light to bring about condensations directly.

CHEMICAL CONDENSATIONS

The general principle of the chemical condensations to be discussed in this section is best illustrated by an example from preparative organic chemistry: the formation of a peptide bond using a carbodiimide such as dicyclohexylcarbodiimide ($C_6H_{11}N{=}C{=}NC_6H_{11}$) as a condensing agent. Such reactions are usually carried out in nonaqueous solvents.

The reaction is believed to occur in two steps.

$$^{+}H_3N-CHR_1-\overset{\overset{\displaystyle O}{\|}}{C}-O^- + RN{=}C{=}NR \xrightarrow{H^+} {}^{+}H_3N-CHR_1-\overset{\overset{\displaystyle O}{\|}}{C}-O-C\begin{array}{l} {\nearrow\!\!\!\!\nearrow} NR \\ \searrow NHR \end{array} \tag{1}$$

$$^{+}H_3N{-}CHR_1{-}\overset{\overset{\displaystyle O}{\|}}{C}{-}O{-}C\begin{matrix}\diagup\!\!\!\!\diagup NR \\ \diagdown NHR\end{matrix} + H_2N{-}CHR_2{-}\overset{\overset{\displaystyle O}{\|}}{C}{-}O^- \rightarrow$$

$$^{+}H_3N{-}CHR_1{-}\overset{\overset{\displaystyle O}{\|}}{C}{-}NH{-}CHR_2{-}\overset{\overset{\displaystyle O}{\|}}{C}{-}O^- + RHN{-}\overset{\overset{\displaystyle O}{\|}}{C}{-}NHR \qquad (2)$$

The overall result is the condensation of two amino acid residues to form a dipeptide, and the hydration of the carbodiimide to a urea by the water generated in the condensation. The hydration of carbodiimides makes available a large amount of free energy. It is this free energy that drives the condensation reaction "uphill." It should be noticed that, since the reaction is usually carried out in a nonaqueous solvent, there is no other source of water except that obtained in the condensation. If the hydration of the carbodiimide is to occur at all, it must be coupled to the formation of a peptide bond. Also, for reasonable yields the amino group should be protected so that it does not react with the carbodiimide.

Carbodiimides and similar reagents may also be used to bring about many of those condensation reactions of phosphate-containing compounds that are important in biochemistry. These include the phosphorylation of nucleosides to give nucleotides, the addition of phosphate to AMP or ADP to give ADP or ATP, respectively, and the formation of internucleotide bonds. A mechanism for one such reaction, the formation of the dinucleotide (APA), may be written

$$\text{Adenosine}{-}O{-}\underset{\underset{\displaystyle O^-}{|}}{\overset{\overset{\displaystyle O}{\|}}{P}}{-}O^- + RN{=}C{=}NR \rightarrow$$

$$\text{Adenosine}{-}O{-}\underset{\underset{\displaystyle O^-}{|}}{\overset{\overset{\displaystyle O}{\|}}{P}}{-}O{-}C\begin{matrix}\diagup\!\!\!\!\diagup NR \\ \diagdown NHR\end{matrix} \qquad (3)$$

$$\big\downarrow \text{Adenosine}$$

$$\text{Adenosine}{-}O{-}\underset{\underset{\displaystyle O^-}{|}}{\overset{\overset{\displaystyle O}{\|}}{P}}{-}O{-}\text{Adenosine} + RHN{-}\overset{\overset{\displaystyle O}{\|}}{C}{-}NHR$$

Models of prebiotic synthesis which attempt to use these reactions must provide the answer to two critical questions. First, how could a suitable condensing agent have originated on the primitive earth and, second, how could such reactions have occurred under mild conditions in an aqueous environment? We shall attempt to answer these questions in turn, but it must be admitted from the outset that we cannot provide a completely satisfactory answer to either.

The type of condensing agent which we are considering in this section is necessarily an unsaturated molecule that releases a substantial amount of energy on hydration. Activated acetylenes such as methoxyacetylene ($CH_3{-}O{-}C{\equiv}CH$) and trichloromethylacetylene ($Cl_3C{-}C{\equiv}CH$) have been used as organic reagents, but cannot plausibly be proposed as prebiotic molecules. On the other hand, cyanoacetylene is formed from methane and nitrogen in an electric discharge and will act as a condensing agent for the formation of organic phosphates.

$$HC{\equiv}C{-}C{\equiv}N + HPO_4^- \rightarrow {}^-HO_3P{-}O{-}CH{=}CH{-}C{\equiv}N$$

$${}^-HO_3{-}P{-}O{-}CH{=}CH{-}C{\equiv}N + ROH \rightarrow$$

$$RO{-}PO_3H^- + H{-}\overset{\overset{\displaystyle O}{\|}}{C}{-}CH_2{-}C{\equiv}N$$

We doubt that cyanamide could have been formed in sufficient quantity to have played a central role in biochemical evolution.

The reagents which have been studied most extensively—all containing a reactive CN triple bond—are cyanamide, cyanamide dimer (cyanoguanidine), and cyanogen. Cyanate, which reacts in the form of cyanic acid, is also a condensing agent.

$H_2N{-}C{\equiv}N$	$(HN{=})(H_2N{-})C{-}NH{-}C{\equiv}N$	$N{\equiv}C{-}C{\equiv}N$	$HN{=}C{=}O$
Cyanamide	**Cyanoguanidine**	**Cyanogen**	**Cyanic acid**

Hydrogen cyanide—which is formed quite readily in electric discharges and has a large free energy of hydrolysis—has not yet been used successfully as a condensing agent, despite a great deal of effort. The tetramer of hydrogen cyanide, however, has been used with some success. A number of other simple unsaturated molecules are known to function well in these reactions but have not been studied in any detail. These include the isonitriles ($RN{\equiv}C$), ketene ($CH_2{=}C{=}O$), and dicyanamide ($N{\equiv}C{-}NH{-}C{\equiv}N$).

Cyanamide is obtained by the ultraviolet irradiation of aqueous solutions of ammonium cyanide. This reaction is inefficient unless certain other ions are present. In the presence of iron, the ferrocyanide that is formed absorbs at longer wavelengths and hence permits more efficient utilization of solar energy. Halide ions perform the same function. The mechanisms of these reactions are unclear, but both must involve radicals formed by charge-transfer processes. It is clear, therefore, that if the oceans once contained appreciable quantities of cyanide and ammonia in addition to ferrous or halide ions, cyanamide could have formed in very large amounts.

The only plausible route to cyanoguanidine is the dimerization of cyanamide, a reaction that goes readily in mildly alkaline solutions.

$$H_2N{-}C{\equiv}N + {}^{-}HN{-}C{\equiv}N \rightarrow H_2N{-}\overset{\overset{\displaystyle N^-}{\|}}{C}{-}NH{-}C{\equiv}N \xrightarrow{H^+} H_2N{-}\overset{\overset{\displaystyle NH}{\|}}{C}{-}NH{-}C{\equiv}N$$

It is not clear that the cyanamide concentration in the ocean or in lakes could ever have been sufficient to bring about this bimolecular reaction, since cyanamide hydrolyzes reasonably quickly, particularly in the presence of inorganic phosphate or phosphate esters.

Cyanogen is obtained both from hydrogen cyanide in an electric discharge and by the pyrolysis of mixtures containing, for example, acetylene and hydrogen cyanide. It is produced by the photochemical oxidation of cyanide ion, particularly in the presence of certain metal ions. Thus, cyanogen is a very plausible prebiotic molecule. Finally, cyanate is obtained in almost quantitative yield by the hydrolysis of cyanogen in neutral or weakly alkaline solution.

$$N{\equiv}C{-}C{\equiv}N + H_2O \rightarrow H_2N{-}\overset{\overset{\displaystyle O}{\|}}{C}{-}C{\equiv}N \rightarrow {}^{-}N{=}C{=}O + CN^- + 2H^+$$

The central difficulty for all theories of prebiotic condensing reactions is the ubiquity of water (not surprisingly, water always interferes with dehydration reactions). Each step in the various condensations must compete with a direct hydration reaction, that is, an uncoupled reaction which uses up condensing agent without producing anything useful; for example,

$$H_2N{-}C{\equiv}N \xrightarrow{H_2O} H_2N{-}\overset{\overset{\displaystyle O}{\|}}{C}{-}NH_2 \quad \text{(Wasteful)}$$

$$H_2N{-}C{\equiv}N \xrightarrow{HPO_4^=} H_2N{-}\overset{\overset{\displaystyle NH}{\|}}{C}{-}O{-}PO_3^= \quad \text{(Potentially useful)} \tag{5}$$

$$H_2N{-}\overset{\overset{\displaystyle NH}{\|}}{C}{-}O{-}PO_3^= \xrightarrow{H_2O} H_2N{-}\overset{\overset{\displaystyle O}{\|}}{C}{-}NH_2 + HPO_4^= \quad \text{(Wasteful)}$$

$$H_2N{-}\overset{\overset{\displaystyle NH}{\|}}{C}{-}O{-}PO_3^= \xrightarrow{ROH} H_2N{-}\overset{\overset{\displaystyle O}{\|}}{C}{-}NH_2 + RO{-}PO_3^= \quad \text{(Useful)} \tag{6}$$

The competition of water in reaction (5) is not too serious since anions usually attack condensing agents much faster than does water. In 0.1 M solutions of inorganic phosphate or phosphate esters, for example, nearly all of the condensing agent undergoes anion addition rather than water addition. It is worth noticing that the activation of inorganic phosphate presents a special problem on account of the low solubility of calcium phosphate. There is very little soluble phosphate in contemporary surface water ($\leq 3 \times 10^{-6}$ M), and the phosphate concentration in the primitive oceans would not have been higher at equilibrium if the pH and Ca^{++} concentrations were the same as now. However, Mg^{+2} inhibits the precipitation of calcium phosphate (hydroxyl apatite), so in the absence of biological precipitation the concentration of phosphate in the ocean and tidepools may have been much greater than the equilibrium value.

The competition of water in reaction (6) is always serious—particularly so in the formation of phosphate esters. Activated phosphates do not discriminate well between alcohols and water, so that even in 1.0 M solutions of nucleosides, for example, this reaction is only about 2% efficient.

Cyanamide and its dimer have also been used to bring about the formation of dipeptides from amino acids, to phosphorylate nucleosides such as uridine, and to form internucleotide bonds. These reactions have usually been carried out in acid solution (pH 2), since the active species in each case is a cation.

$$^{+}H_3N-C\equiv N \qquad H_2N-\overset{\overset{\displaystyle ^{+}NH_2}{\|}}{C}-NH-C\equiv N$$

Such low pH's are unlikely to have occurred, except in very special environments on the primitive earth. However, we do not think this objection too serious, since at higher pH's the same reactions do occur with cyanamide, although much more slowly. We do not know how efficient such syntheses would be at these higher pH's.

Cyanogen is much more reactive than cyanamide or its dimer. It brings about the phosphorylation of nucleosides quite rapidly, although inefficiently, at 0°C and pH 6 to 8. The phosphorylation of adenosine, uridine, and cytidine has also been demonstrated using cyanogen generated photochemically from ferricyanide, or ferrocyanide and oxygen.

Cyanate has not been used successfully either in the formation of peptide or phosphodiester bonds, although organic isocyanates are often used as condensing reagents in synthetic chemistry. Cyanate has been used in one prebiotic reaction, the synthesis of pyrophosphate from orthophosphate on the surface of hydroxylapatite. Since the formation of carbamyl phosphate from cyanate and phosphate in neutral aqueous solution is well known

$$HN=C=O + HPO_4^{=} \rightleftarrows H_2N-\overset{\overset{\displaystyle O}{\|}}{C}-O-PO_3^{=}$$

it is presumed that carbamyl phosphate is an intermediate and reacts with a second molecule of phosphate on the surface of the hydroxylapatite.

$$H_2N-\overset{\overset{\displaystyle O}{\|}}{C}-O-PO_3^{=} + HPO_4^{=} \longrightarrow H_2P_2O_7^{=} + NH_4^{+} + HCO_3^{-}$$

It is interesting to note that carbamyl phosphate can react with ADP to give ATP, in the presence of a suitable enzyme. It is also an essential intermediate in pyrimidine and arginine biosynthesis. Thus, carbamyl phosphate is a high-energy phosphate, can be formed from inorganic materials under very mild conditions, and is used in the biosynthesis of components of proteins and nucleic acids. It may well have been an important compound in the evolution of the genetic apparatus.

THERMAL SYNTHESIS

Condensation reactions which lead to the formation of polypeptides and polynucleotides from their components can be brought about by the heating of dry mixtures. Nonetheless, it is by no means clear which of these reactions should be considered as potentially prebiotic.

The phosphorylation of nucleosides can be achieved by heating them with acid phosphates such as NaH_2PO_4 and $Ca(H_2PO_4)_2$. The reaction occurs quite rapidly at 130°C and can still be studied at temperatures as low as 65–80°C. The formation of internucleotide bonds occurs under similar conditions. A particularly striking case is the formation of di and trinucleotides by heating uridine 2′-3′-cyclic phosphate with uridine at 130°C. The yield of the trinucleotide was as high as 7%. Furthermore, the most abundant isomer of the trinucleotide was 3′-5′ linked.

There are two reasons for doubting the relevance of these observations to problems of prebiotic chemistry. First, it seems unlikely that the pH of either the oceans or the larger lakes on the primitive earth could ever have been low enough to permit the formation of acid salts of phosphoric acid, such as $Ca(H_2PO_4)_2$ or even $Ca(HPO_4)$. Neutral or alkaline salts such as hydroxylapatite do not take part in these phosphorylation reactions. This criticism also applies to all schemes involving the reactions of the free acid forms of nucleotides or of cyclic nucleotides, and even more strongly to reactions carried out in anhydrous phosphoric acid or polyphosphoric acid.

The second objection is quite different. The formation of phosphate ester bonds is accompanied by many side reactions. Among these, the dephosphorylation of nucleotides is important, and this would not permit extensive polymerization. It would, however, permit the conversion of a fraction of the nucleoside present to nucleotides and short oligonucleotides.

It is possible that an "acid" environment for phosphate condensations could come about indirectly. If a solution containing a nucleoside and ammonium phosphate is evaporated down and then heated, ammonia is lost and $(NH_4)H_2PO_4$ left behind. Phosphorylation reactions take place in these mixtures. The formation of ammonium phosphate by evaporation of sea water is extremely unlikely, since magnesium ammonium phosphate would deposit first. We shall see that a mixture of magnesium ammonium phosphate and urea is an excellent phosphorylating agent. Ammonium phosphate (or the phosphate of an organic base) might have been formed by the evaporation of a lake on the primitive earth.

A related system consisting of urea, ammonium chloride, and an inorganic phosphate is a much more active phosphorylating medium. In a typical experiment, one equivalent each of uridine and sodium hydrogen phosphate, and ten equivalents each of urea and ammonium chloride were dissolved in water and adjusted to pH 8.5 with ammonia. The solution was evaporated to dryness and heated at 100°C. After 24 hours, more than 90 % of the phosphate had been incorporated into nucleotides. If the temperature was maintained at 65°C, the reaction was slower, but after 22 days a yield of about 40 % of nucleotides was obtained. If hydroxylapatite was used in place of sodium phosphate, a yield of 20–25 % of nucleotides was obtained after 24 hours at 100°C.

The first products formed in these reactions are nucleoside-2′- and 3′-phosphates, and nucleoside-5′-phosphates, but the former condense a second time to give the 2′-3′-cyclic phosphates. The course of a typical reaction is illustrated in Fig. 11-1. In the absence of a nucleoside, condensed inorganic phosphates are obtained in good yield at 100°C.

It is generally believed that hydroxylapatite was common on the primitive earth since it is the most abundant source of phosphate on earth today. Urea is formed in many prebiotic reaction sequences. Thus, the system of ammonium chloride, urea, and hydroxylapatite is quite plausible as a prebiotic phosphorylating mixture. The urea can be replaced by many substituted ureas and some simple amides, but it is not clear that these compounds would have been abundant on the primitive earth.

The Mg^{++} ion has a specific catalytic effect in the presence of urea and inorganic phosphate, for it strongly enhances the formation of pyrophosphate bonds. If 5′-uridylic acid is heated with ammonium dihydrogen phosphate at 65–85°C the main products are a mixture of diphosphates such as uridine-3′-5′-diphosphate. In the presence of Mg^{++}, this reaction is suppressed and a mixture of UDP and the pyrophosphate UppU is obtained instead. This provides a plausible mechanism for the formation of pyrophosphate bonds on the primitive earth.

We conclude that the phosphorylation of nucleosides and the formation of short oligonucleotides may have occurred by purely thermal processes. Indeed, such compounds could have been important as starting materials for further syntheses in aqueous solution. However, we doubt that very extensive polymerization of nucleotides could have occurred in this way, or that "biological" polymerization could have taken place except in an aqueous environment.

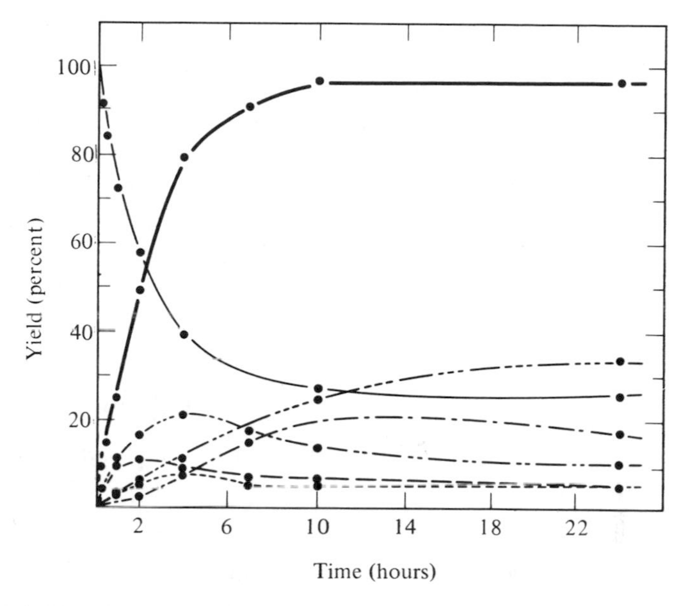

Fig. 11-1. Phosphorylation of uridine with a mixture of inorganic phosphate, ammonium chloride, ammonium bicarbonate, and urea at 100°C. U, ———; Up!, – – – –; pU, — · — ·; Up, — — — —; pUp!, — · · —; pUp - - - - -; total incorporation of inorganic phosphate, ———. U = Uridine; Up! = Uridine-2′,3′ cyclic phosphate; pU = Uridine-5′-phosphate; Up = Uridine-3′-phosphate; pUp! = Uridine-5′-phosphate-2′,3′-cyclic phosphate; pUp = Uridine-3′,5′ diphosphate. (From Lohrman and Orgel, *Science* **171**, 494, 1971.)

THERMAL SYNTHESIS OF POLYPEPTIDES

The thermal synthesis of polypeptides has been studied very extensively, particularly by S. W. Fox. When a pure amino acid is heated to about 180°C, tarry products, varying quantities of diketopiperazines, and small amounts of polypeptides are formed. However, if a mixture of amino acids

```
          H
          N
RHC ⁄        ⁀ C=O
   |          |
O=C          CHR
       ⁀ N ⁄
          H
```

Diketopiperazine

rich in acidic or basic amino acids is used, substantial yields of polypeptides can be obtained at 150–180°C. The molecular weights of the products can be as high as 20,000 when glutamic acid is heated with glycine, for example. If a mixture of the 20 natural amino acids is used in this experiment, all of them are incorporated into polypeptide-like materials, as would be expected.

These materials have been called "proteinoids," but we prefer to refer to them as *thermal polypeptides.*

Most of the bonds in thermal polypeptides are of the normal type, but those formed from mixtures which include aspartic and glutamic acid also contain α-β and α-γ amide links, respectively. Abnormal bonds are also formed to the ε-amino group of lysine. It has been claimed that the sequences of the polypeptides formed in these polymerizations are not random, but the evidence on this point is inadequate. Many of the polypeptide chains begin with glutamic acid; this may be due to the formation of pyroglutamic acid from glutamic acid and the subsequent initiation of the peptide chains by the pyroglutamic acid.

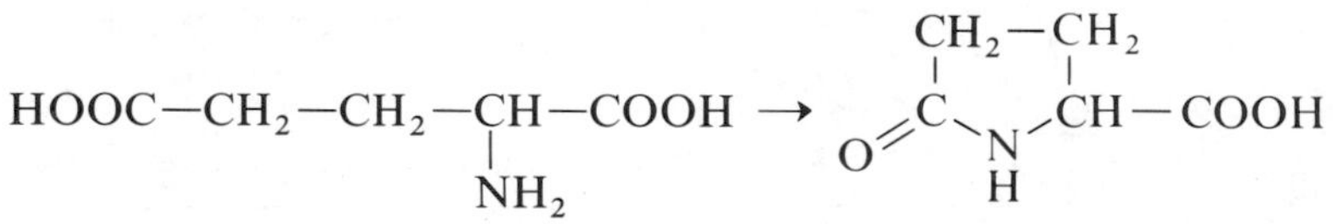

$$HOOC{-}CH_2{-}CH_2{-}\underset{\displaystyle NH_2}{\underset{|}{CH}}{-}COOH \longrightarrow \text{(ring: } CH_2{-}CH_2 \text{ bridging } O{=}C \text{ and } CH{-}COOH \text{ via } N{-}H\text{)}$$

There is no evidence to show whether the amino acids within a chain are highly ordered or not.*

If the product formed by heating a mixture of amino acids is boiled with water, a dispersion of microspheres is obtained. *Microspheres* are more or less spherical colloidal particles, often with diameters close to 2μ. Some microspheres are surrounded by structures that look like bilayer membranes. They are not "biological-like" membranes, since they do not contain lipids or carry out any of the functions of biological membranes. Some microspheres, however, look very much like simple bacteria under a microscope. Under appropriate conditions, microspheres can be made to divide into two or to form buds. Similar behavior has been known in other colloidal systems for a long time and a possible relation between the division of colloidal particles and cell division has been commented on by a number of authors. It seems unlikely, in the light of recently acquired information about molecular genetics, that the division of microspheres is related to the origin of cell division (see Chapter 12).

Thermal polypeptides display a catalytic activity in some reactions that is greater than that of the amino acids which they contain. However, they are less active than a number of small molecules. The reactions which have been reported to be catalyzed by random polypeptides are (so far) ones which go rather easily without polypeptide catalysts (e.g., the hydrolysis of p-nitrophenyl acetate). It seems possible that simple peptides exist that catalyze

* The probability of obtaining a specified polypeptide containing 100 amino acids in a sample of random polypeptides is 20^{-100}. The chance of obtaining a peptide with glutamic acid at the end, on the other hand, is 20^{-1}. Thus, the degree of nonrandomness in thermal polypeptides so far demonstrated is minute compared with the nonrandomness of proteins. It is deceptive, then, to suggest that thermal polypeptides are similar to proteins in their nonrandomness.

such important synthetic reactions as, for example, internucleotide-bond formation. The thermal synthesis of polypeptide catalysts for a reaction of this kind would mark an important advance in our understanding of prebiotic chemistry.

The importance of these thermal syntheses in prebiotic chemistry is a very controversial matter. We do not believe that they were very important because we doubt that polypeptides could have been synthesized in large quantities at the surface of the earth by thermal reactions of the kind so far demonstrated. If a mixture of amino acids were buried several kilometers beneath the surface of the earth, the temperature would be high enough so that thermal polypeptides might be formed; nevertheless, they would be turned to tar in the thousands or millions of years it would take to erode away the overlying material. In some hot springs the temperature may rise to 150–180°C. However, such hot springs, which are not abundant, are not suitable places for the thermal synthesis of polypeptides, which requires dry conditions. Polymerization of amino acids does not take place in the presence of liquid water at 150–180°C; indeed, the presence of liquid water would at these temperatures hydrolyze polypeptides to amino acids. The only other places where temperatures of 150–180°C are attained are in volcanoes; and here the conditions for synthesis are not favorable. In a volcano, the temperature of the molten lava is about 1200°C, a temperature that will pyrolyze the amino acids completely. When the lava solidifies, the surface of the lava is close to the temperature of the air, although it is hot beneath the surface. Except for holes or cracks in the cooling lava which might get hot enough, a volcano is not a suitable place to conduct a thermal synthesis of polypeptides.

Another way of examining this problem is by asking whether there are places on the earth today with appropriate temperatures where we could drop, say, 10 grams of a mixture of amino acids, and obtain a significant yield of polypeptides. (The course of this reaction would not be affected by oxidizing or reducing conditions in the atmosphere.) We cannot think of a single such place.* Thus, we believe that thermal syntheses of polypeptides at these high temperatures did not take place on an extensive scale† on the primitive earth. If the thermal synthesis of polypeptides was important for the origin of life, it probably occurred at more moderate temperatures.

* If there were places where such a polymerization could be accomplished, then it would still be necessary to show how the amino acids were brought to the lava and the peptides removed from it in an efficient manner.

† The reader can probably construct a scenario where such a polymerization could have taken place a few times in geological history. However, such a rare process would not be an important prebiotic synthesis. An example of such a rare process is the natural occurrence (other than in meteorites) of metallic iron on the earth. This iron is formed when a lava flow intrudes into a coal bed. The iron oxide in the lava is reduced to metallic iron by the time the lava has cooled. There are only a few places where such iron has been found.

OTHER CONDENSATION MECHANISMS

(1) Inorganic trimetaphosphate takes part in two reactions that are potentially prebiotic. It reacts with *cis*-glycols to give phosphates. With nucleosides, excellent yields of the 2′- and 3′-phosphates can be obtained at room temperatures and pH's in excess of 10.

$$\text{HO—OH}_2\text{(ribose; Base; OH, OH)} + \text{trimetaphosphate} \longrightarrow \text{HOCH}_2\text{(ribose; Base; 2',3'-cyclic phosphate } O\text{—P(=O)O}^-\text{—O)} + P_2O_7^{4-}$$

The reaction occurs more slowly at neutral pH's and the terminal yield is then decreased owing to the slow acid-catalyzed hydrolysis of the phosphates.

Trimetaphosphate brings about the formation of dipeptides in up to 30% yield and small amounts of tripeptides from simple amino acids at alkaline pH's. The reaction proceeds via a cyclic intermediate.

Both of these reactions are highly specific and produce useful products. If a plausible source of trimetaphosphate could be suggested, they would merit serious consideration as prebiotic reactions.

(2) Activated phosphates can be generated by the oxidation of a quinol phosphate to give a highly active phosphorylating species, presumably the phosphate of a quinol radical or quinone. Many biochemical phosphorylating agents are generated by oxidation: diphosphoglyceric acid and reduced nicotinamide dinucleotide are generated from glyceraldehyde phosphate, inorganic phosphate, and the oxidized form of nicotinamide dinucleotide, for example. So far there is no evidence that such processes could occur under prebiotic conditions, but the possibility should be borne in mind in view of their great importance in contemporary metabolism.

(3) A number of mechanisms involving the formation of polypeptides from amino acid precursors have been proposed; from the amino acid nitriles, for example.

$$n(\text{H}_2\text{N—CHR—CN}) \longrightarrow \text{H}_2\text{N—CHR—}\overset{\text{NH}}{\overset{\|}{\text{C}}}\text{—NH—CHR—}\overset{\text{NH}}{\overset{\|}{\text{C}}}\text{—NH}\ldots \longrightarrow$$

$$\text{H}_2\text{N—CHR—}\overset{\text{O}}{\overset{\|}{\text{C}}}\text{—NH—CHR—}\overset{\text{O}}{\overset{\|}{\text{C}}}\text{—NH}\ldots$$

In no case does it appear likely that such reactions could have occurred prebiotically.

(4) The aminoacylimidazoles give polypeptides in aqueous solution.

$$^{+}H_3N{-}CHR{-}\overset{\overset{\displaystyle O}{\|}}{C}{-}N\langle\text{imidazole}\rangle N + H_2N{-}CHR{-}\overset{\overset{\displaystyle O}{\|}}{C}{-}O^{-} \rightarrow$$

$$^{+}H_3N{-}CHR{-}\overset{\overset{\displaystyle O}{\|}}{C}{-}NH{-}CHR{-}\overset{\overset{\displaystyle O}{\|}}{C}{-}O^{-} + HN\langle\text{imidazole}\rangle N$$

A prebiotic synthesis of these or related aminoacyl imidazolides would be important.

(5) Direct photochemical polymerization reactions have occasionally been discussed. As far as we are aware they have not been achieved as yet under prebiotic conditions.

REACTIONS OF THE AMINOACYL PHOSPHATES ON CLAYS

One quite novel synthesis of polypeptides has recently been reported. The aminoacyl phosphates are highly activated compounds (the anhydrides of amino acids and phosphoric acid derivatives). They react in aqueous solution to give short peptides in high yield; for example,

$$n\left(H_3N^{+}{-}CH_2{-}\overset{\overset{\displaystyle O}{\|}}{C}{-}O{-}PO_3{-}\text{adenosine}\right) \rightarrow$$

$$H_3N^{+}{-}CH_2{-}\overset{\overset{\displaystyle O}{\|}}{C}{-}\left(NH{-}CH_2{-}\overset{\overset{\displaystyle O}{\|}}{C}{-}\right)_{n-2}{-}NH{-}CH_2{-}\overset{\overset{\displaystyle O}{\|}}{C}{-}O^{-} + n\text{AMP}$$

Even in the absence of catalysts or adsorbents, this is one of the most efficient syntheses of peptides in aqueous solution.

Recently it has been discovered that certain layered clays, the Montmorillonites, adsorb very large amounts of the aminoacyl adenylates between their layers and then cause extensive polymerization to a discrete series of long oligomers. In one experiment, ala_{30}, ala_{32}, ala_{42}, and ala_{56}, along with shorter peptides, were obtained from alanyl adenylate. The efficiency of this striking reaction is almost 100%.

These observations are particularly exciting because, as we saw in Chapter 6, the aminoacyl adenylates are the substrates of protein synthesis. A plausible prebiotic synthesis of these anhydrides would constitute a major advance in prebiotic chemistry.

This chapter has probably been confusing to the reader. We believe that this is because of the very limited progress that has been made in the study of prebiotic condensation reactions. Many interesting scraps of information are available, but there is no indication that the correct pathways have yet been discovered. We have gone into considerable detail because we think this is one of the most important problems in prebiotic chemistry and one in which considerable progress can be anticipated in the near future.

References

Cyanamide

G. D. Steinman, R. M. Lemmon, and M. Calvin, "Cyanamide: a possible key compound in chemical evolution," *Proc. Nat. Acad. Sci. U.S.* **52**, 27 (1964).

R. Lohrmann and L. E. Orgel, "Prebiotic synthesis: phosphorylation in aqueous solution," *Science* **161**, 64 (1968).

C. Ponnamperuma and E. Peterson, "Peptide synthesis from amino acids in aqueous solution," *Science* **147**, 1572 (1965).

J. D. Ibanez, A. P. Kimball, and J. Oró, "Possible prebiotic condensation of mononucleotides by cyanamide," *Science* **173**, 444 (1971).

Cyanamide dimer and dicyanamide

G. D. Steinman, D. H. Kenyon, and M. Calvin, "Dehydration condensation in aqueous solution," *Nature* **206**, 707 (1965).

G. D. Steinman, D. H. Kenyon, and M. Calvin, "The mechanism and photobiochemical relevance of dicyanamide mediated peptide synthesis," *Biochim Biophys. Acta* **124**, 339 (1966).

G. D. Steinman, R. M. Lemmon, and M. Calvin, "Dicyandiamide: possible role in peptide synthesis during chemical evolution," *Science* **147**, 1574 (1965).

Cyanogen and cyanoformamide

R. Lohrmann and L. E. Orgel, "Prebiotic synthesis: phosphorylation in aqueous solution," *Science* **161**, 64 (1968).

H. Halman, R. A. Sanchez, and L. E. Orgel, "Phosphorylation of D-ribose in aqueous solution," *J. Org. Chem.* **34**, 3702 (1969).

H. Halman and H. L. Schmidt, "Cyanogen-induced synthesis of ^{18}D-labelled β-ribofuranase-1-phosphate and its acid catalyzed hydrolysis," *J. Chem. Soc. (C)* **1970**, 1191.

Cyanate

S. L. Miller and M. Parris, "Synthesis of pyrophosphate under primitive earth conditions," *Nature* **204**, 1248 (1964).

R. Lohrmann and L. E. Orgel, "Prebiotic synthesis: phosphorylation in aqueous solution," *Science* **161**, 64 (1968).

Cyanoacetylene

J. P. Ferris, "Cyanovinyl phosphate: a prebiological phosphorylating agent," *Science* **161**, 53 (1968).

J. P. Ferris, G. Goldstein, and D. J. Beaulieu, "An evaluation of cyanovinyl phosphate as a prebiotic phosphorylating agent," *J. Am. Chem. Soc.* **92**, 6598 (1970).

HCN tetramer

S. Chang, J. Flores, and C. Ponnamperuma, "Peptide formation mediated by hydrogen cyanide tetramer: a possible prebiotic process," *Proc. Nat. Acad. Sci. U.S.* **64**, 1011 (1969).

Ethyl metaphosphate and polyphosphoric acid

G. Schramm, H. Grotsch, and W. Pollmann, "Non-enzymatic synthesis of polysaccharides, nucleosides and nucleic acids and the origin of self-reproducing systems," *Angew. Chem. Int. Ed.* **1**, 1 (1962). These syntheses use ethyl metaphosphate, a reagent prepared from phosphorus pentoxide, diethyl ether, and chloroform, which is not a prebiotic reagent. The synthesis of nucleosides has been reproduced by other workers, but the template syntheses described have not.

T. V. Waenheldt and S. W. Fox, "Phosphorylation of nucleosides with polyphosphoric acid," *Biochim. Biophys. Acta* **134**, 1 (1967). We do not regard polyphosphoric acid as a prebiotic reagent.

A. Schwartz and S. W. Fox, "Thermal synthesis of internucleotide phosphodiester linkages," *Biochim. Biophys. Acta* **87**, 696 (1964).

A Schwartz and S. W. Fox, "Condensation of cytidylic acid in the presence of polyphosphoric acid," *Biochim. Biophys. Acta* **134**, 9 (1967).

Polyphosphates and metaphosphates in aqueous solution

J. Rabinowitz, S. Chang, and C. Ponnamperuma, "Phosphorylation by way of inorganic phosphate as a potential prebiotic process," *Nature* **218**, 442 (1968).

A. Schwartz and C. Ponnamperuma, "Phosphorylation of adenosine with linear polyphosphate salts in aqueous solution," *Nature* **218**, 443 (1968).

W. Feldmann, *Z. für Chem.* **9**, 154 (1969). Peptide synthesis using trimetaphosphate.

J. Rabinowitz, J. Flores, R. Krebsbach, and G. Rogers, "Peptide formation in the presence of linear or cyclic polyphosphates," *Nature* **224**, 795 (1969).

J. Rabinowitz, "Condensation reactions with linear and cyclic polyphosphates" *Helv. Chim. Acta* **52**, 2663 (1969).

J. Rabinowitz, "Peptide and amide bond formation in aqueous solution of cyclic or linear polyphosphates as a possible prebiotic process," *Helv. Chim. Acta* **53**, 1350 (1970).

N. Chung, R. Lohrmann, L. E. Orgel, and J. Rabinowitz, "The mechanism of the trimetaphosphate-induced peptide synthesis," *Tetrahedron* **27**, 1205 (1971).

Phosphorylation by heating dry mixtures

S. Chang, C. Ponnamperuma, and J. Rabinowitz, "Phosphorylation by way of inorganic phosphate as a potential prebiotic process," *Nature* **218**, 442 (1968).

C. Ponnamperuma and R. Mack, "Nucleotide synthesis under possible primitive earth conditions," *Science* **148**, 1221 (1965).

R. Lohrmann and L. E. Orgel, "Urea-inorganic phosphate mixtures as prebiotic phosphorylating agents," *Science* **171**, 490 (1971).

Effect of Mg^{++} ions on the solubility of hydroxyl apatite

G. J. Handschuh and L. E. Orgel, "Struvite and prebiotic phosphorylation," *Science* **179**, 483 (1973).

C. S. Martens and R. C. Harris, "Inhibition of apatite precipitation in the marine environment by magnesium ions," *Geochim. Cosmochim. Acta* **34**, 621 (1970).

Thermal polypeptides

S. W. Fox and K. Harada, "The thermal copolymerization of amino acids common to protein," *J. Amer. Chem. Soc.* **82**, 3745 (1960).

S. W. Fox and K. Harada, "Thermal copolymerization of amino acids in the presence of phosphoric acid," *Arch. Biochem. Biophys.* **86**, 281 (1960).

S. W. Fox, K. Harada, K. R. Woods, and C. R. Windsor, "Amino acids compositions of proteinoids," *Arch. Biochem. Biophys.* **102**, 439 (1963).

S. W. Fox and K. Harada, "Characterization of thermal polymers of neutral alpha-amino acids with dicarboxylic amino acids of lysine," *Arch. Biochem. Biophys.* **109**, 49 (1965).

S. W. Fox and T. Nakashima, "Fractionation and characterization of an amidated thermal 1:1:1-proteinoid," *Biochim. Biophys. Acta* **140**, 155 (1967).

S. W. Fox and T. V. Waehneldt, "The thermal synthesis of neutral and basic proteinoids," *Biochim. Biophys. Acta* **160**, 246 (1968).

G. Biserte and P. A. Finot, "Polycondensation thermique des alpha-amino-acides (Cinétique de la reaction de Fox and Harada)," in *Biogenèse: Colloque sur les Systèmes Biologiques Élémentaires et la Biogenèse* (Mason, Paris, 1967), pp. 278–300.

G. Hennon, R. Plaquet, M. Dautrevaux, and G. Biserte, "Synthèse de polymères d'acides aminés par polycondensation thermique et étude de quelques caractères physicochimiques," *Biochemie* **53**, 215 (1971).

F. Hare, "Ètude cinètique de la polycondensation thermique d'α-amino acides," *J. Chim. Phys.* **68**, 330 (1971).

Microspheres from thermal polypeptides

S. W. Fox, K. Harada, and J. Kendrick, "Production of spherules from synthetic proteinoid and hot water," *Science* **129**, 1221 (1959).

S. W. Fox, "How did life begin?" *Science* **132**, 200 (1960).

S. W. Fox and S. Yuyama, "Dynamic phenomena in microspheres from thermal proteinoid," *Comp. Biochem. Physiol.* **11**, 317 (1964).

S. W. Fox, R. J. McCauley, and A. Wood, "A model of primitive heterotrophic proliferation," *Comp. Biochem. Physiol.* **20**, 773 (1967).

Catalytic activities

S. W. Fox and G. Krampitz, "Catalytic decomposition of glucose in aqueous solution by thermal proteinoids," *Nature* **203**, 1362 (1964).

D. L. Rohlfing and S. W. Fox, "The inactivation of catalytically active thermal polyanhydro-alpha-amino acids," *Arch. Biochem. Biophys.* **118**, 127 (1967).

H. G. Hardebeck, G. Krampitz, and L. Wulf, "Decarboxylation of pyruvic acid in aqueous solution by thermal proteinoids," *Arch. Biochem. Biophys.* **123**, 72 (1968).

K. Dose and L. Zaki, "Recent progress in the study and abiotic production of catalytically active polymers of α-amino acids," in *Chemical Evolution*, R. Buvet and C. Ponnamperuma, eds. (North-Holland, Amsterdam, 1971), p. 263.

Peptide synthesis from amino acid adenylates on clays

M. Paecht-Horowitz, J. Berger, and A. Katchalsky, "Prebiotic synthesis of polypeptides by heterogeneous polycondensation of amino acid adenylates," *Nature* **228**, 636 (1970).

Chapter Twelve

From Random Polymers to the Most Primitive Organism

INTRODUCTION

In the preceding chapters we concentrated on the prebiotic synthesis of amino acids and nucleotides and on the condensation of these molecules to give polypeptides and polynucleotides. It has often been argued that such an approach is too narrow since it presupposes a close resemblance between the earliest forms of life and present day organisms. We believe that this criticism, while tenable from a purely logical point of view, is completely sterile unless accompanied by alternative proposals concerning the nature of the primitive system. Such proposals, if they are to be taken seriously, should specify how the alternative genetic system worked and how it evolved into the nucleic acid-protein system. Even more important, theories should be subject to some kind of experimental test. The suggestion that the first genetic system was made up of specific carbohydrates, for example, is too indefinite to be tested. This criticism applies even more strongly to theories in which silicon-containing compounds play a key role.

Why do we believe that the first organisms were dependent on polypeptides and polynucleotides? The strongest argument has already been given, namely that amino acids, sugars, purines, and pyrimidines are major products of reactions under prebiotic conditions. They are not the exclusive products, but they are the major products—more often than could be reasonably expected on grounds of chance alone. The simplest explanation of this coincidence is that amino acids, sugars, and so on, were incorporated into the first organisms because they were abundant on the primitive earth.

The theory that the first organisms were similar to contemporary organisms is also attractive because it is the simplest theory. Any other has to explain the evolution of the nucleic acid-protein system from some other, unrelated system. While it should be remembered that there are logically sound alternatives, the most natural first objective of *experiments* on the origins of life is to find out whether or not the present system could have evolved without systems having a completely different biochemistry preceding it. At present, there is no reason to believe that it could not.

For these reasons, we shall consider the emergence of the first organism only within the framework of the nucleic acid-protein system. We shall discuss three specific proposals.

1. The first form of life was based on proteins but did not utilize nucleic acids.
2. The first form of life involved nucleic acid replication but did not use the nucleic acids to code for protein synthesis.
3. The first form of life was based on nucleic acid replication and used nucleic acids to code for protein synthesis.

LIFE BASED ON PROTEINS ALONE

The spontaneous formation of polypeptides seems, at first sight, more likely than the prebiotic synthesis of nucleic acids because:

1. In laboratory experiments, amino acids are produced from mixtures of simple gases under less restrictive conditions than are the nucleotide bases.
2. The sugar components of the nucleic acids—ribose and deoxyribose—are much less stable in aqueous solution than are the amino acids. It is, therefore, harder to understand how they could have accumulated prebiotically.
3. The formation of the purine nucleosides has been demonstrated under "prebiotic" conditions, but the synthesis is not completely satisfactory.
4. The condensation of α-amino acids to polypeptides is unambiguous, except for aspartic acid, glutamic acid, and lysine. The 3′–5′ linkage of the nucleic acid backbone, on the other hand, represents only one of a number of alternative ways of joining together nucleotides (3′–3′, 5′–5′, for example). Thus we might expect the polymerization of nucleotides to nucleic acids under prebiotic conditions to be much less efficient than the polymerization of amino acids.

It seems likely, therefore, that the nonbiological polymerization of amino acids preceded, or was contemporary with, the appearance of nucleic acids. It then seems natural to ask, how far could a system of proteins without nucleic acids have developed towards life as we know it? It appears to us that no great progress would have been possible in the absence of a fairly accurate form of residue-by-residue replication (see Chapter 6).

Suppose first, that the primitive soup contained a variety of amino acids and an energy source to bring about their condensation to polypeptides. The first polymers obtained would have constituted a very complex mixture. The sequences would not have been completely random since some amino acids react faster than others and since the presence of a particular amino acid at the end of a growing chain might have had some influence on the nature of the next amino acid incorporated. However, from what we know about the polymerization of mixtures of amino acids, it seems that the degree of ordering would not have been very great. In the course of time the least stable peptide bonds would have hydrolyzed. Finally a steady state would have been reached in which synthesis and hydrolysis were balanced.

We do not understand the properties of random polypeptides sufficiently to predict with confidence the properties of the steady-state mixture, but it seems at least possible that some of the components would have possessed catalytic activity. If so, these components might have influenced the relative abundance of the different amino acids in the peptide and also the nearest-neighbor frequencies in the steady-state mixture of peptides. At first sight it might seem that if these catalytic effects were important, the distribution of polypeptides in the steady state might have corresponded to a high level of biological organization.

We believe that this conclusion is incorrect because it fails to recognize an intrinsic difference between "statistical" methods of developing and maintaining biological order and residue-by-residue replication. If a sufficiently accurate process of residue-by-residue replication exists, then once a single molecule of a given peptide sequence is formed it can give rise to any number of progeny protein molecules with the same sequence. Thus, although any individual molecule must ultimately be degraded, the effect of a single polymer may persist indefinitely. Furthermore, if replication is sufficiently but not completely accurate, then, under suitable conditions, evolution by mutation and natural selection is possible, just as for nucleic acids. For reasons given below, we do not believe that "statistical" models can be formulated that have these properties, and hence we do not believe that models proposing cycles of chemical reactions are likely to correspond to advanced stages in the origin of life.

The production of a well-defined and fairly long polypeptide sequence, except by residue-by-residue replication, requires a new "enzyme" for each kind of link. However, each enzyme is itself a well-defined and fairly long polypeptide chain, and so in turn requires an entire further series of enzymes. It seems most unlikely that a self-consistent replicating system of this type could exist, and even less likely that it would be capable of evolving. We are thus forced to consider systems in which long polypeptides are not reproduced accurately but in which strong statistical requirements are imposed on the sequences generated. Preformed peptides must modify the probability with which the different amino acids occur as neighbors in newly formed chains.

In attempting to design a self-replicating system of this kind we are faced with a serious dilemma. If the restrictions on nearest neighbors are weak, then no order can be maintained for more than a very few residues; if they are strong, the variety of polypeptides which are formed is strictly limited, for once a few residues are connected together the rest of the sequence is determined. Thus, we see that such a system is incapable of evolving by making a few discrete changes in the sequence of otherwise fully determined chains; all that can happen is that the entire pattern of nearest-neighbor probabilities may change and so cause complicated changes in the nearest-neighbor distribution of all polypeptides in the system.

If we accept this conclusion it follows that a highly organized biological system based on proteins is possible if, and only if, a process of unit-by-unit replication of polypeptides could have developed spontaneously.

A mechanism involving complementary replication comparable to polynucleotide replication is perhaps most plausible. A sequence of positively charged, uncharged, and negatively charged residues could replicate, with positively charged and negatively charged amino acids pairing with each other and neutral amino acids going together, as shown in Fig. 12-1. This would give three-component, partially complementary, replicating sequences. A two-letter system, based on asparagine-glutamine pairing, or a four-letter system, based on lysine-aspartic acid and arginine-glutamic acid pairing, might have a stereochemical basis.

While replication schemes of this kind are by no means excluded, we are prejudiced against them. We doubt whether, in fact, they could be realized using any of the amino acids known to be synthesized in prebiotic experiments. Furthermore, we do not see how the transition from self-replicating

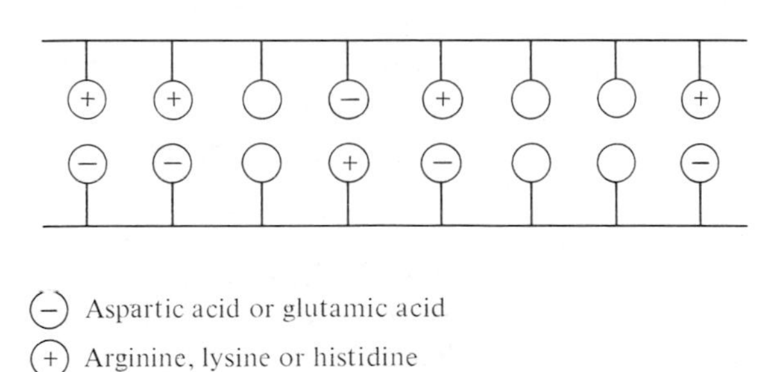

Fig. 12-1. A hypothetical pairing scheme for polypeptides. Positive charges pair with negative charges and aliphatic side chains are self-pairing by means of hydrophobic interactions.

proteins to the nucleic acid-protein system could have occurred. These objections are not overwhelming; experimental attempts to realize such systems might be rewarding.

A second mode of residue-by-residue replication must also be considered. Could a copying enzyme (or series of enzymes) which brought about residue-by-residue replication have come into existence by chance, even though, spontaneously, there was no tendency for such replication? While we do not seriously question that a copying system of this kind is in principle possible, we believe that the obstacles to its evolution are enormous, because no mode of natural selection exists in the early stages of its development.

We do not wish to imply that amino acids and polypeptides played no role in prebiotic chemistry prior to the development of the modern protein-synthesizing system, but it does seem unlikely that a self-replicating polypeptide system could have evolved. Even if it could, we still find it difficult to understand how such a system would have developed into the present nucleic acid-protein system.

The conclusion of this section is that life based on proteins alone is possible only if random polypeptides made up from some small set of amino acids had the capacity for identical or complementary replication. We do not think this likely, but we admit that it is a possibility worthy of experimental study.

LIFE BASED ON NUCLEIC ACIDS WITHOUT A PROTEIN CODE

The basic genetic process is the replication of nucleic acid, DNA or RNA. To decide whether or not life based on nucleic acids alone might have evolved to any considerable degree of complexity in the absence of a well-defined relation between nucleic acids and protein synthesis, we must consider two problems: replication without enzymes and evolution without proteins.

We shall argue that the complementary replication of nucleic acids makes use of certain structural characteristics inherent in the bases themselves. Thus, it is quite plausible that reasonably accurate replication is possible in the absence of proteins.

The evidence for this belief comes from a variety of studies of the physical chemistry of nucleotide bases and their derivatives. Using the bases and their simple derivatives, it has been possible to prepare mixed crystals containing, for example, 7-methyladenine and 4-methyluracil, or 7-methylguanine and 4-methylcytosine. X-ray studies show that the crystals contain well-defined dimers which in some, but not all, cases are hydrogen-bonded, as in the complementary base-pairs of DNA. Studies of solutions of the bases in organic solvents are even more suggestive. Self-complexing and complexing between all possible pairs of the naturally occurring bases (A, G, C, and

U) have been investigated. It has been found that the A-U and G-C pairs associate more strongly than any others.

We do not know whether there are sets of molecules significantly different from the standard base pairs which could form the basis of a genetic system. It is easy to write down new complementary base pairs by modifying nonessential parts of the standard bases. Barbituric acid, for example, could substitute for uracil, or 8-azaguanine for guanine. More drastic changes, such as the replacement of purines by corresponding tricyclic derivatives or changes in the nucleic acid backbone itself, also seem possible—on paper.

Unfortunately, we do not know which, if any, of these systems would work in practice. Nor can we tell whether the components of these systems could have been synthesized and would have accumulated in sufficient quantities on the primitive earth. Thus, we cannot say whether the nucleic acids are the only potential genetic molecules likely to arise spontaneously on a primitive planet, or whether many other equally satisfactory systems are possible. When we know more about the details of the origins of life on earth we shall probably be in a position to deal more satisfactorily with this question.

TEMPLATE SYNTHESIS

The interactions between synthetic polynucleotides in aqueous solution have been studied extensively. It has been shown, for example, that polyuridylic acid and polyadenylic acid form a double-helical structure similar to that of DNA. In addition, they form a triple-helix, poly U + poly U + poly A, in which hydrogen-bonded base triplets occur. Poly C + poly C + poly G forms an analogous helical structure (Fig. 12-2).

The results of studies on the interaction of mononucleotides and short oligonucleotides with polynucleotides are more interesting. Polyuridylic acid forms stable triple helices, 2 poly U-X, with many monomeric adenosine derivatives—with adenosine, adenosine-5′-phosphate and adenosine 2′-3′-cyclic phosphate, for example. Double-helices are stable only if the amino group of adenosine is substituted; 6-methylaminopurine, for example, forms a double helix with poly U. Polycytidylic acid forms triple-helices with many guanosine derivatives, but polycytidylic acid also forms double helices readily. By contrast, no organized helical structures are formed by polyadenylic acid or polyguanylic acid with complementary monomeric pyrimidine nucleosides or their derivatives.

Monomeric purine nucleoside derivatives, when incorporated into helical structures with complementary polymers, can be made to undergo efficient condensation to di- and oligonucleotides. A number of experiments using nucleotides and a condensing agent such as a carbodiimide have been reported. Alternatively, preactivated nucleotides such as the N-phosphoimidazolides have been used directly.

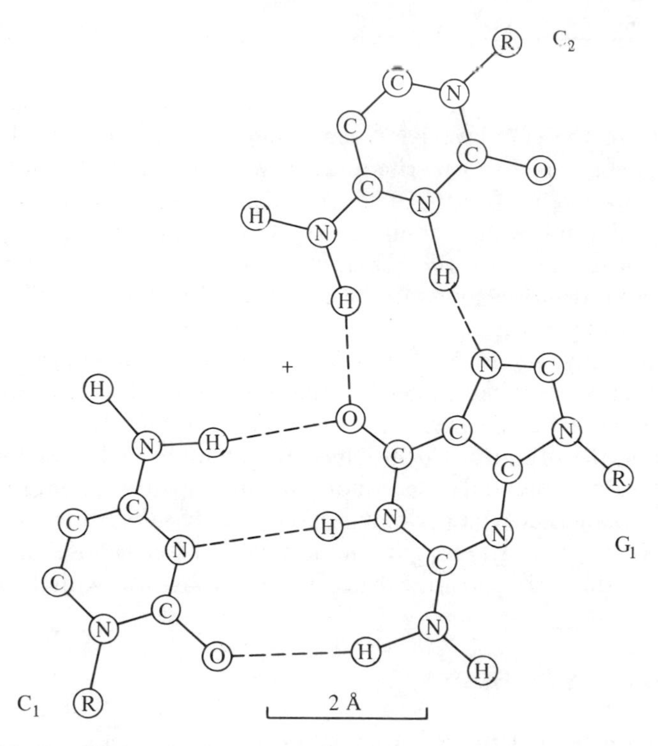

Fig. 12-2. Possible bonding schemes between G and C in a three-strand helix formed between 5′-GMP and 2 poly C. (Redrawn from *Biochem. Biophys. Res. Comm.* **17**, 93, 1964.)

At temperatures below the melting point of the poly U + poly U + pA triple helix (that is, when the bases are stacked), polyuridylic acid greatly increases the yield of di- and short oligoadenylic acids obtained from adenylic acid and adenosine in the presence of a water-soluble carbodiimide. On the other hand, it has no effect on the combination of adenylic acid with uridine, cytidine, or guanosine. In a reciprocal fashion, poly C facilitates the condensation of pairs of guanosine monomers, but has no effect on the condensation of guanylic acid with uridine, cytidine, or adenosine. Finally, poly U has no effect on the condensation of guanylic acid with guanosine, and poly C has no effect on the condensation of adenylic acid with adenosine. Thus the Watson-Crick pairing rules, which are the key to nucleic acid replication, apply to these simple chemical syntheses.

Pyrimidine monomers do not form organized structures with poly A or poly G under any of the conditions tested. There is also good reason to believe that such structures cannot exist, except at temperatures very much below 0°C. Consequently, attempted template-directed syntheses using pyrimidine monomers have not been successful. This difficulty could perhaps be overcome by using preformed dinucleotides or oligonucleotides. Alternatively, it may be possible to incorporate pyrimidines between pairs of purines.

Water-soluble carbodiimides are, of course, not prebiotic molecules. Cyanamide, a prebiotic molecule closely related to the carbodiimides, cannot easily be used in template syntheses since it reacts so slowly at the low temperatures (20°C) at which the organized helices are stable. Cyanogen, which is much more reactive, has been tried, but did not give any product. However, a related molecule, cyanogen chloride, works well in this system. It is not clear whether cyanogen chloride is a prebiotic compound.

The reactions of preactivated nucleotides are more interesting from the point of view of chemical evolution. While ATP does form a stable complex with poly U in the presence of Mg^{++}, no condensation occurs; the ATP slowly hydrolyzes to ADP and then to AMP. The imidazolides of the nucleotides, on the other hand, are excellent reagents. Adenosine-5′-phosphoimidazolide (ImpA), for example, reacts in better than 50% yield on a poly U template to give short oligomers. It is interesting to note that the active intermediates in the reactions of several enzymes that transfer phosphate groups from ATP have been shown to be N-phospho derivatives of histidine.

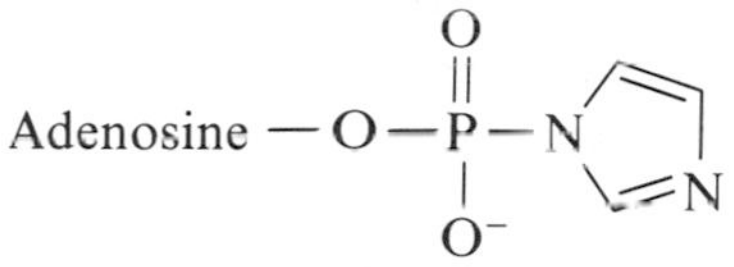

$$\text{Adenosine}-O-\overset{\overset{\displaystyle O}{\|}}{\underset{\underset{\displaystyle O^-}{|}}{P}}-N\langle\text{imidazole}\rangle$$

ImpA

More recently it has been found that although adenosine 2′-3′-cyclic phosphate (Ap!) does not undergo self-condensation on a poly U template in the presence of Mg^{++} as a stabilizing cation, it does give dinucleotides and trinucleotides in good yield if certain simple catalysts replace the Mg^{++} ion. Glycinamide, ethylenediamine, or propylenediamine, for example, are effective.

All natural nucleic acids are 3′–5′ linked, but in the template-directed condensation, the 2′–5′ linkage predominates. With carbodiimide, a great deal of 5′–5′ dinucleotide is also formed, but in many reactions of Ap! and ImpA, almost pure 2′–5′ linked compounds are obtained. Perhaps mixed 2′–5′ and 3′–5′ linked oligonucleotides were important in the early phases of chemical evolution, and the 2′–5′ linkages were eliminated only after the development of efficient polymerases.

The template-directed reactions are sensitive to the nature of the sugar component. Either ribonucleotides or deoxyribonucleotides may be used to provide the activated phosphate group, but ribonucleosides are much better acceptors than deoxynucleosides, presumably because a conveniently oriented 2′-OH group is present. Deoxyadenosine is a moderate acceptor, while arabinosyl adenine and α-adenosine are poor acceptors. Polyuridylic acid (D form) is a very effective template for the reaction of ImpA (D form) with D-adenosine but is a poor template (<10%) for the reaction of L-adenosine.

These findings are relevant to the nature of the first nucleic acid. Arguments can be presented in favor either of DNA or of RNA. Deoxyribonucleosides, for example, are less stable to acid hydrolysis than ribonucleosides, while ribose can be synthesized more easily than deoxyribose under prebiotic conditions. Template reactions go better with ribose than deoxyribose, but polyribonucleotides are less stable than polydeoxynucleotides to alkaline hydrolysis, and so on. We doubt that it is worth trying to decide this issue, since the first nucleic acids may not have been as homogeneous as the compounds synthesized enzymatically. The first nucleic acids may have contained many different sugars, including perhaps ribose, deoxyribose, arabinose, and even certain hexose sugars.

The formation of the initial polymers may have taken place on a surface. This would have had the dual role of concentrating monomers from dilute solution and perhaps also of directing, by a noncomplementary process, the initial synthesis of partly ordered polymers. The surface might have been inorganic in origin (e.g., a mineral), or organic (e.g., a mixture of polypeptides which had arisen independently of the polynucleotide system).

Let us accept, tentatively, the conclusion that reasonably accurate replication of polynucleotides using mononucleotides as precursors may have occurred in a nonbiological environment. What degree of chemical organization would then have been possible without the coupling of nucleic acids and protein synthesis?

It seems clear that if complementary template-directed synthesis is sufficiently accurate and sufficiently rapid compared with competing random polymerization processes, then a given ancestral polynucleotide may give rise to any number of descendants. Furthermore, a certain degree of natural selection might operate: Under any given set of conditions, those sequences which replicate most rapidly would tend to outgrow all others. In practice, given inaccurate replication coupled with degradative reactions, we would not expect families of identical molecules, but rather spatially confined families of related polymer molecules having similar but not identical sequences.

If certain sequences replicate more rapidly than others over a wide range of conditions, then they might become dominant over geographically extended regions. We do not yet know enough about these reactions to say whether this is likely, but certainly we cannot exclude the possibility that a homopolymer or some simple repeating sequence would in fact win out in such a competition. Thus, if only A and U were present, AUAUAU... might become the dominant sequence, with occasional errors permitting AUAUAAUAU..., and so forth.

What is needed now to bring this process nearer to contemporary life? The missing feature is the development of polymers which modify the chemistry of their environment in ways other than by acting as direct templates for complementary synthesis. These polymers might, for example:

1. Increase the rate of template replication of other polynucleotides in their environment (primitive nucleotide polymerases).
2. Increase the supply of activated precursors in their own environment (primitive synthetic "enzymes").
3. Destroy unrelated polynucleotides (primitive nucleases).

At present there is no evidence that polynucleotides have even limited catalytic activity. In biological systems, we know that catalytic functions are performed by proteins and never by polynucleotides. However, it is not impossible that polynucleotides, perhaps with the help of cofactors, are able to catalyze chemical reactions, and that this ability was important during the evolution of life. Their catalytic function would subsequently have been taken over by the much more versatile polypeptides. Thus, the question of the catalytic activity of polynucleotides remains open.

It is worth remembering that our overall view of the structure of nucleic acids is dominated by the work of Watson and Crick, who showed that DNA has a regular structure with the bases unexposed. More recent work on the base sequence of transfer RNAs makes it seem likely that precisely defined structures of great complexity can be built using only the standard nucleotides.

It seems to us that a polynucleotide chain with well-defined secondary structures might be able to bind amino acids strongly and specifically. This would be important for the evolution of the genetic code. We doubt whether polynucleotides could exhibit strong catalytic activity, although we cannot exclude this possibility. Our overall conclusion is that the development of life probably awaited the emergence of a primitive genetic code.

LIFE BASED ON NUCLEIC ACIDS AND PROTEINS

We have seen that "organisms" without nucleic acids would lack the means of achieving genetic continuity, while organisms without proteins would be severely limited in their ability to use the chemicals in their environment. The difficulties associated with theories of the direct evolution of life as we know it are of a quite different kind. While the efficiency of organisms using both nucleic acids and proteins is not in doubt, we do not understand how they could have evolved. In particular, we do not understand the origin of the genetic code that provides the critical connection between the genetic and the functional apparatus of the cell.

Nowadays, the stage in which amino acids are first specifically associated with polynucleotides is the loading of the t-RNAs. This step is brought about by a set of activating enzymes, each of which recognizes a specific amino acid and its corresponding t-RNA. Once an amino acid is attached to its t-RNA, it is, in most cases, transferred to a growing protein chain without further discrimination.

It seems unlikely that there is a direct and specific interaction between amino acids and their t-RNAs. Nonetheless, it is clear that at some early stage in the evolution of life the direct association of amino acids with polynucleotides, which was later to evolve into the genetic code, must have begun. At this time no activating enzymes could have been in existence, if by enzymes we mean catalytically active polypeptides produced under the direction of polynucleotides. It is difficult to avoid the conclusion that the original association was achieved by direct selective interactions between amino acids and polynucleotides, possibly assisted by polypeptides or other organic material which had been formed previously. Theories of the origin of the code differ in many aspects, but particularly in the assumptions they make concerning the nature of this selective interaction.

The simplest theory suggests that the role of t-RNA was originally filled by a set of much shorter polynucleotides, perhaps the anticodon trinucleotides. In this form, the theory postulates that trinucleotides have a selective affinity for the amino acid coded by their complementary trinucleotide. Of course, the selectivity must have been limited in the first place, but it is argued that it might have been sufficient to produce primitive activating enzymes in the presence of a suitable messenger RNA. Once this was done, the system could have perfected itself. If this type of theory is correct, the code is not arbitrary; if life were to start again, certain features of the code would be reproduced because it would be based on the same physical-chemical interactions.

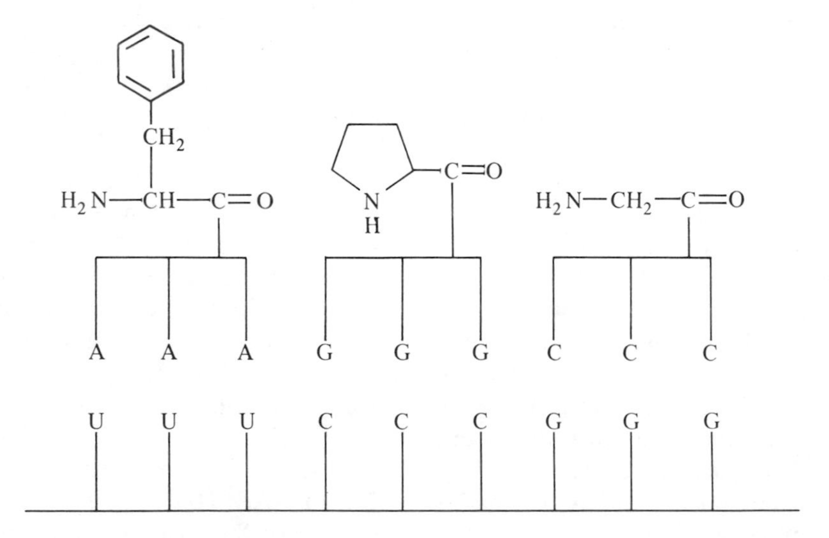

Fig. 12-3. Hypothetical scheme for protein synthesis in which an amino acid is attached to the anticodon trinucleotide.

There are two difficulties with these theories. Firstly, the anticodons are too short to have well-defined secondary structures. It does not seem likely that systematic discrimination between amino acids could be achieved by complexing with trinucleotides. Secondly, the amino acids attached to trinucleotides would be held far apart in space by a polynucleotide template (Fig. 12-3). It seems unlikely that they could condense together efficiently. These objections are not conclusive; experimental work to test theories postulating specific amino acid-anticodon interaction is in progress in a number of laboratories.

The second approach to this problem separates the codon-anticodon interaction from the polynucleotide-amino acid recognition from the beginning. We suppose that primitive but fairly complex t-RNAs were a feature of the translation mechanism from the start. In order for a polynucleotide to act as a primitive t-RNA it would require at least an anticodon loop and a region which interacted specifically with one or more of the available amino acids. The structure should also allow the attachment of the selected amino acid to the 3′-OH group of the terminal nucleotide. Finally, the secondary structure of the t-RNA should be such that if two t-RNAs are held to adjacent positions on a polynucleotide messenger, their amino acids are brought close enough together to permit peptide formation.

THE EVOLUTION OF THE DETAILS OF THE GENETIC CODE

The genetic code is given in Table 6-1. The facts about the code which need to be explained by any reasonably complete theory are:

1. There are four different bases in messenger RNA.
2. Each codon consists of a sequence of three bases.
3. There are 20 amino acids, although a three letter code could, in principle, permit up to 64 of them.
4. The code is the same in all organisms.
5. xyU and xyC *always* code for the same amino acid.
6. xyA and xyG *usually* code for the same amino acid.
7. In eight cases, xyU, xyC, xyA, and xyG all code for the same amino acid.
8. In most cases, all codons for a single amino acid start with the same pair of bases. Exceptions are leucine, serine, and arginine.
9. There are three triplets, UAA, UAG, and UGA, that signal the end of a protein chain.
10. The triplet AUG is sometimes a special signal to start a new protein chain. As a chain initiator it codes for formylmethionine rather than methionine in *E. coli.*
11. The allocation of amino acids to positions in the code is not random.

All codons with U in the central position correspond to hydrophobic amino acids. Charged amino acids always are coded for by triplets with a purine in the second position. A number of other regularities have been noted.

A number of authors have tried to account for these facts, but none has given a theory which we regard as satisfactory. Hence we can only give our own opinions on some of the main points that have been made.

There is no way of deciding whether the code has always involved the same four bases or whether the primitive code had only two. We are reasonably confident that codons have always consisted of three letters, for there is no easy way of going from a one or two letter code to a three letter code. However, it is quite possible that only one or two of the letters of the primitive codons were significant.

We think it unlikely that a stereospecific relationship existed between codons or anticodons and corresponding amino acids. It is more likely that t-RNAs are primitive, and that amino acids were recognized by "pockets" in the t-RNAs. If this is correct the genetic code is a "frozen accident"; once a fairly complex code and a large set of proteins had evolved, any change would have inactivated so many enzymes that it would have proved lethal. If a code were to evolve independently, for example on another planet, we do not know whether it would be similar to the code on the earth.

It is naïve to believe that the code in the early stages of its evolution was as precise as it is now. Presumably, all the amino acids in the environment, including both D- and L-amino acids, were able to be incorporated into polypeptides, as were other related molecules (such as simple carboxylic acids and amines). Again it seems likely that each codon specified a group of related amino acids rather than a single amino acid. It is hard to believe that valine could be distinguished from isoleucine or serine from threonine, for example, until the code had become highly developed.

We clearly do not understand how the code originated. New ideas that can be tested experimentally are needed.

References

Template syntheses

J. Sulston, R. Lohrmann, L. E. Orgel, and H. T. Miles, "Nonenzymatic synthesis of oligoadenylates on a polyuridylic acid template," *Proc. Nat. Acad. Sci. U.S.* **59**, 726 (1968).

J. Sulston, R. Lohrmann, L. E. Orgel, and H. T. Miles, "Specificity of oligonucleotide synthesis directed by polyuridylic acid," *Proc. Nat. Acad. Sci. U.S.* **60**, 409 (1968).

B. J. Weimann, R. Lohrmann, L. E. Orgel, H. Schneider-Bernloehr, and J. E. Sulston, "Template-directed synthesis with adenosine-5′-phosphoimidazolide," *Science* **161**, 387 (1968).

Interesting theories of the origins of protein synthesis and the genetic system are given in:

F. Lipmann, "Gramicidin S and tyrocidine biosynthesis: A primitive process of sequential addition of amino acids on polyenzymes," in *Chemical Evolution and the Origin of Life*, R. Buvet and C. Ponnamperuma, eds. (North Holland, Amsterdam, 1971), pp. 381–391; *Science* **173**, 875 (1971).

A. G. Cairns-Smith, *The Life Puzzle* (University of Toronto Press, Toronto, 1971).

E. D. Hanson, "Evolution of the cell from promordial living systems," *Quart. Rev. Biol.* **41**, 1 (1966).

An interesting theoretical discussion of evolution, natural selection at the molecular level, and the origin of life is given by M. Eigen, *Naturwissenschaften* **58**, 465–523 (1971), and is recommended to the mathematically inclined reader. A less mathematical version of this paper is in *Quart. Rev. Biophys.* **4**, 149 (1971).

H. Kuhn, "Self-organization of molecular systems and evolution of the genetic apparatus," *Angew. Chem. Int. Ed.* **11**, 798 (1972).

Extended discussions of the genetic code are given in:

The Genetic Code, Cold Spring Harbor Symposium on Quantitative Biology, Volume 31 (1966).

C. R. Woese, *The Genetic Code* (Harper & Row, New York, 1967).

T. H. Jukes, *Molecules and Evolution* (Columbia University Press, New York, 1966).

C. T. Caskey, "The universal genetic code," *Quart. Rev. Biophys.* **3**, 295 (1970).

Chapter Thirteen

Optical Activity

The presence of optically active organic compounds in living organisms is one of their most striking characteristics. The proteins are composed entirely of L-amino acids (and glycine, which has no asymmetric carbon). The nucleic acids contain only D-ribose and D-deoxyribose. D-glucose plays a central role in carbohydrate metabolism, but L-glucose is unimportant. Since Pasteur's discovery of optically active compounds, these facts have had a curious fascination for many writers on the origins of life. Pasteur himself wrote "The universe itself is a disymmetrical assemblage. I am inclined to think that life, as it appears to us, must be a product of the disymmetry of the universe. . . ."

In our later discussion we shall need the following elementary facts about optically active molecules.

(1) *The direction in which the plane of polarization of light is rotated by a substance is indicated by the symbols d and l, or* (+) *and* (−). D and L, as applied to amino acids, do not indicate the direction of this rotation, but

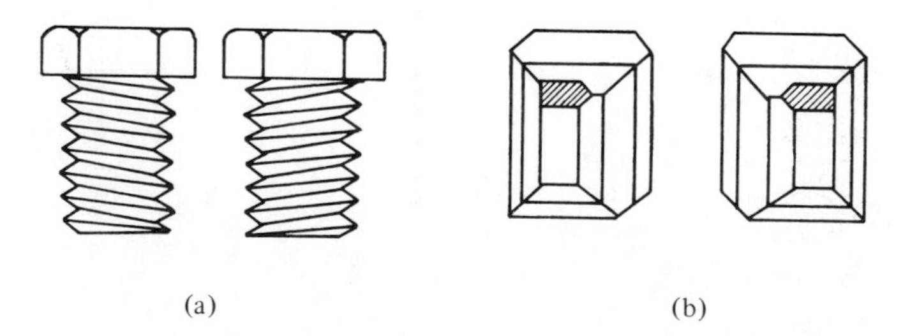

Fig. 13-1. Two examples of objects whose mirror images will not superimpose. (a) Left-handed and right-handed screws. (b) D- and L-forms of sodium ammonium tartrate. Pasteur separated crystals of sodium ammonium tartrate by hand in 1848 to achieve the first resolution of an optically active compound.

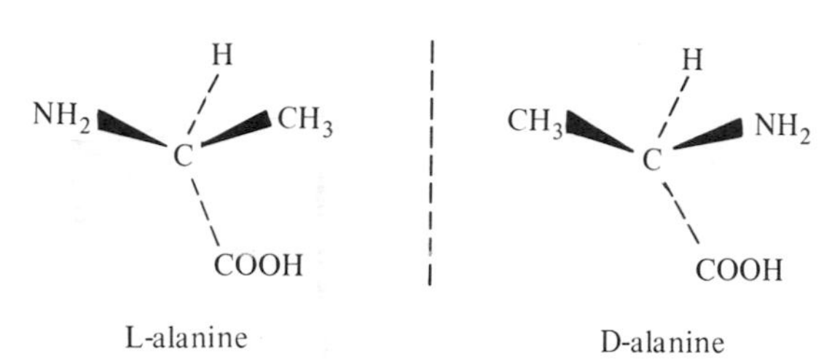

Fig. 13-2. L- and D-alanine showing that they are mirror images. If L-alanine is twisted around 180°, it can not be superimposed on D-alanine.

rather show the configuration of the substituents at the α-carbon (see Fig. 13-2). D and L in the sugar series refer to the configuration of groups at the 4-carbon in pentoses and the 5-carbon in hexoses (Fig. 13-3). The configuration of the D amino acids and the D sugars are related to each other by an arbitrary procedure. One must not try to give a deep meaning to the appearance of "D" before sugars but "L" before amino acids.

(2) *The stabilities of the two optical enantiomers of a compound are the same in an optically inactive environment.* Thus, the free energy of D-alanine and L-alanine are identical in free space or in pure water. This rule applies to systems of molecules, however complex. The mirror image of the reader would have the same stability as the reader himself. However, this rule does not apply in an optically active environment—for example, D- and L-alanine have different free energies of solution in D-sec butanol.

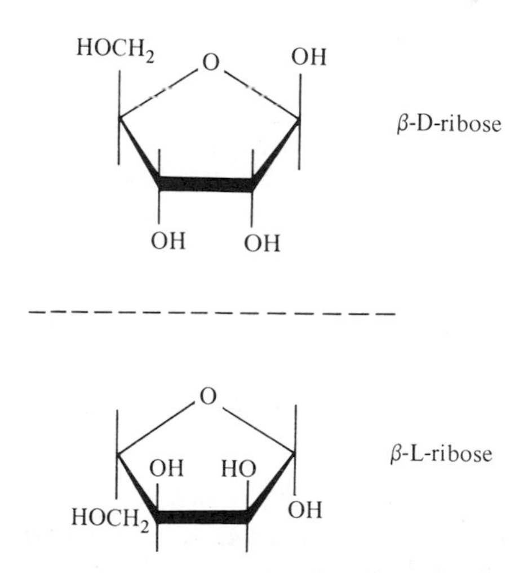

Fig. 13-3. The optical isomers of ribose. The symbols D and L refer to the absolute configuration at the 4-carbon atom (5-carbon atom in the case of hexoses). Once the configuration at the 4-carbon is determined, the configuration at the other carbon atoms is fixed, since each hydroxyl group occupies a known position (*cis* or *trans*) relative to the 4-carbon.

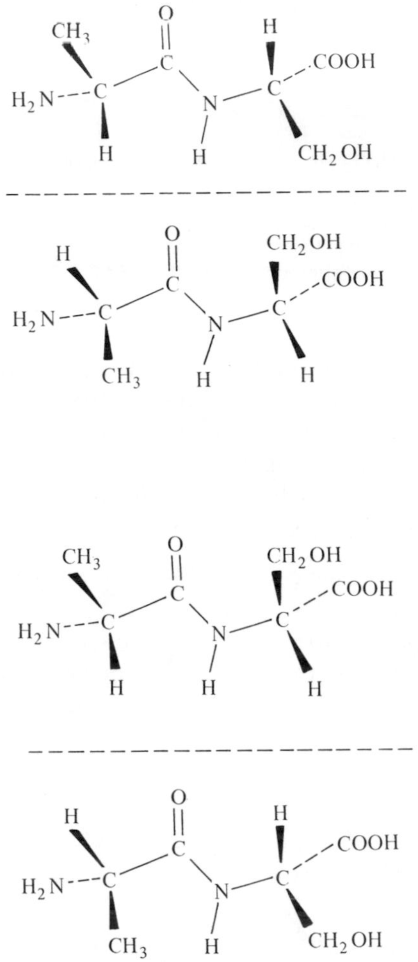

(a) L-alanyl-D-serine

(b) D-alanyl-L-serine

(c) L-alanyl-L-serine

(d) D-alanyl-D-serine

Fig. 13-4. The four isomers of alanyl-serine. The dotted line is the mirror plane. (a) and (b) are mirror images, as are (c) and (d). (a) and (c) or (d) are different compounds with different properties.

(3) *The rates of reaction of D and L isomers with an optically inactive system are the same.*

(4) *The interaction of two optically active molecules is dependent, sometimes strongly, on their relative configurations.* If we start with two racemic compounds, DL-A and DL-B, we can form two pairs of complexes, D-A-L-B, L-A-D-B and D-A-D-B, L-A-L-B. The members of each pair are optical isomers and so have the same energy. The members of the different pairs (D-A-L-B, L-A-D-B, and D-A-D-B, L-A-L-B) may have very different energies and may be formed or react at quite different rates.

(5) *Two enantiomers may behave as completely different compounds in their interactions with a complex, optically specific system such as an enzyme.* Thus, it is probably easier for an enzyme to distinguish D-phenylalanine from L-phenylalanine than for it to distinguish D-phenylalanine from D-isoleucine.

(6) *Chemical syntheses of compounds capable of optical activity always give a racemic mixture if symmetric reagents are used in the synthesis.* If asymmetric reagents or solvents are used, the product may be optically active, but in most cases the optical activity is barely detectable.

It is often a difficult task to resolve a racemic mixture into its optical isomers. Resolution is usually achieved by reacting DL-A with L-B, forming D-A-L-B and L-A-L-B, which can then, in principle, be separated by fractional crystallization. In practice, this fractional crystallization is frequently unsuccessful because the solubilities of D-A-L-B and L-A-L-B are not sufficiently different.

(7) *An optically inactive mixture, whether of optically active molecules or not, can generate optical activity spontaneously under certain circumstances.* A supersaturated solution of DL-glutamic acid under carefully controlled conditions can precipitate one of the isomers, either through seeding or spontaneously. The remaining solution (mother liquor), is also optically

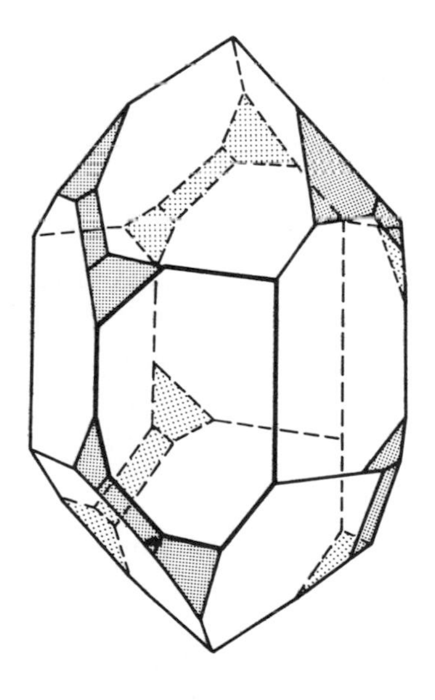

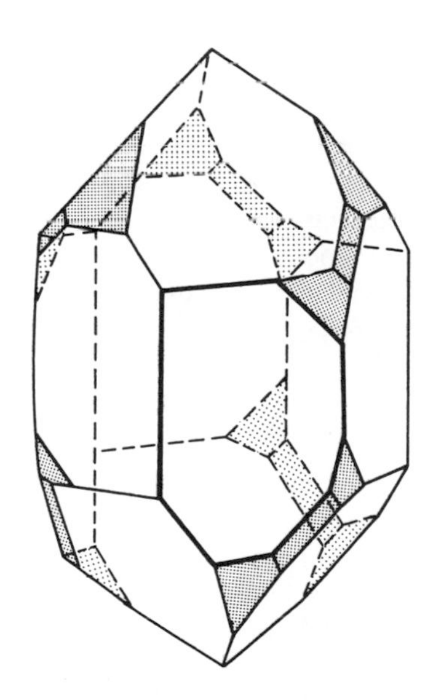

Fig. 13-5. Left-handed and right-handed quartz. Quartz crystals rotate the plane of polarized light because the crystals contain right- and left-handed helical chains of silicate tetrahedra. Individual tetrahedra would not be optically active. Right and left quartz occur in equal abundance in nature.

active because it contains a greater concentration of the isomer that did not crystallize. The Japanese use the seeding technique as a step in their industrial synthesis of monosodium L-glutamate (MSG). The L-glutamic acid is crystallized out from a racemic mixture. The mother liquors contain D-glutamic acid, which is then racemized to DL-glutamic acid, and the procedure repeated. If the spontaneous crystallization without seeding is carried out many times, the number of experiments giving D-glutamic acid will be about the same as the number giving L-glutamic acid.

An example of an optically inactive substance generating optical activity is the precipitation of silica as quartz. The optical activity of quartz crystals is due to the helical arrangement of ... Si—O—Si—O ... chains in the crystal, rather than to the presence of asymmetric molecules. The mother liquors are, of course, optically inactive.

Naturally, we would expect that equal quantities of right- and left-quartz crystals would be formed if this experiment were repeated many times. Large quantities of optically active quartz occur in the earth's crust, and it has been repeatedly suggested that a predominance of one of the optical forms of quartz is connected with the origin of optical activity in living organisms. The abundance of the two forms of quartz in the same quartz bed has been measured. In one count of 6,404 crystals, 50.05% were left quartz, and 49.95% were right quartz; the difference is not statistically significant. Other counts of quartz crystals give comparable results.

(8) *The earth does not provide a totally symmetric environment for organic synthesis.* There is a direction defined by its rotation and, at any point, another defined by the magnetic field. There is also a direction implicit in some nuclear reactions due to the breakdown of parity in weak nuclear interactions. We believe that it is extremely unlikely that any of these asymmetries could have coupled strongly enough with any of the relevant chemical reactions on the primitive earth to have effected a choice of D or L isomers.

Circular polarized light is another source of asymmetry on the earth. This could arise by reflection of plane polarized light in the earth's magnetic field. Circular polarized light has been used to produce optical activity by the selective decomposition of the enantiomers of α-bromopropionic acid ethyl ester and of α-azidopropionic acid dimethyl amide. The percent resolution of these compounds was extremely small, even after 50% of the starting material had been destroyed. It is most unlikely that this effect could have generated significant amounts of optically active compounds on the primitive earth.

(9) *If optically-active compounds were synthesized on the primitive earth, they would have been important only if they were stable for long periods in the ocean or in sediments.* While most optically active hydrocarbons do not racemize, sugars and amino acids do. Optically active amino acids racemize rapidly (on the geological time scale) through the ionization of the α-hydrogen

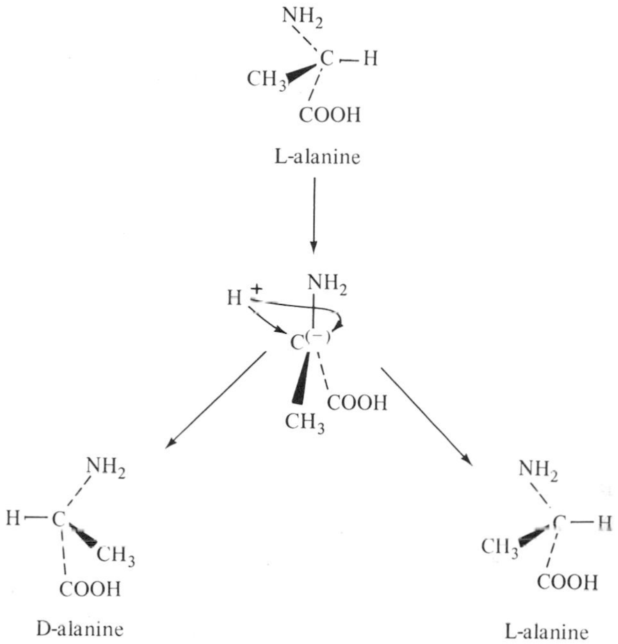

NH_2 — C — H, CH_3, COOH

L-alanine

H^+, NH_2, $C^{(-)}$, COOH, CH_3

NH_2, H — C, CH_3, COOH

D-alanine

NH_2, C — H, CH_3, COOH

L-alanine

Fig. 13-6. Racemization of L-alanine by formation of a carbon ion on the α hydrogen. The carbon ion is planar; the hydrogen ion can add from the right giving L-alanine or from the left giving D-alanine. It is equally probable for the hydrogen to add from the right or from the left.

(Fig. 13-6). The rate of racemization is independent of pH between pH 3 and 8 for isoleucine, alanine, and phenylalanine; the respective half-lives are 4.4×10^6, 1.1×10^6, and 16,000 years at 0°C, and 35,000, 11,000, and 2,000 years at 25°C. The racemization of amino acids in peptides is more rapid at pH's above about 10, because the amide group makes the α-hydrogen substantially more acidic than it is in the dipolar ion of the free amino acid. However, at pH values between 5 and 8, the rate of hydrolysis of the peptide bond is greater than the rate of racemization. The sugars undergo more complex anomerization reactions quite rapidly under mild conditions.

These facts make it clear that the optical rotation of the primitive ocean was zero. Polypeptides containing only L-amino acids could occur, but only along with an equal number of polypeptides containing only D-amino acids. The same would hold for the ribose in the nucleic acids. *It is therefore an accident that living systems on the earth use L-amino acids and D-ribose rather than D-amino acids and L-ribose.*

How did it come about that living organisms incorporated only L-amino acids in their proteins and D-sugars in their nucleic acids? This is not a baffling problem. We can offer a number of possible explanations. We just do not know, as yet, which is the correct one.

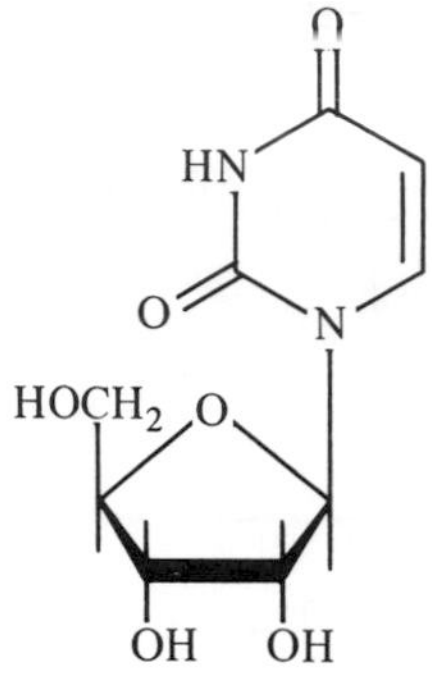

β-D-uridine

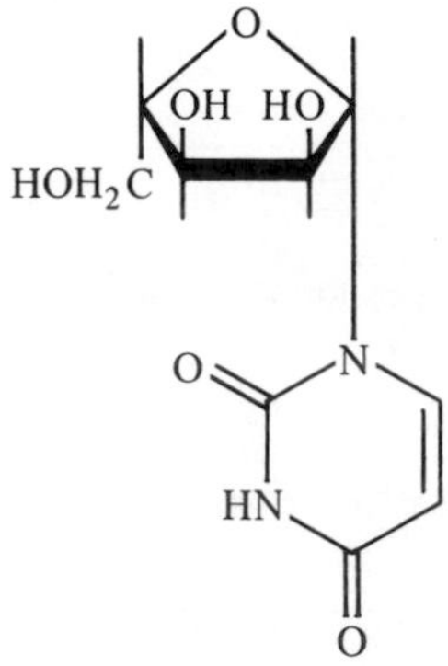

β-L-uridine

Fig. 13-7. The optical isomers of β-uridine. Changing from D- to L-β-uridine involves inversion of each substituent in the plane of the furanose ring. If an attempt is made to insert L-β-uridine into a double stranded nucleic acid helix, the 3′- and 5′-hydroxyls will not join up correctly with the rest of the sugar-phosphate backbone.

(1) The relative configurations must be correct at the 1-, 3-, and 4-carbon atoms of a pentose sugar if the formation of a DNA-like structure is to be possible. It is clear that replicating double-stranded nucleic acids can be built either with all D-ribotides or all L-ribotides. However, it would not be possible to build a regular structure with a mixture of the two. It has been shown experimentally that an activated form of D-adenylic acid will condense with D-adenosine (but not with L-adenosine) on a poly-D-uridylic acid template. Thus the optical homogeneity of the nucleic acids is probably a necessity if they are to function as genetic materials.

(2) There may be a similar direct structural reason for the optical specificity of the proteins, but this is not at all obvious. Certainly, D-amino acids do not fit very well into an α-helix containing mainly L-amino acids, but then proteins are known which contain little or no α-helix. Limited regions of α-helix of either polarity could be formed in "mixed DL" proteins by confining the amino acids in a limited length of polypeptide to those of a single handedness. It is perhaps worth noting that the spontaneous polymerization of certain racemic amino acid derivatives leads to the formation of short stretches of polypeptide containing only D-, or only L-amino acids.

(3) Perhaps the development of the process of protein synthesis is possible only if a fixed relation exists between the enantiomers of the amino acids and nucleotides. We do not know whether peptide-bond formation takes place close enough to the polynucleotide messenger for one group of molecules to influence the other directly. Even if this is not the case now, it may have been for the first protein-synthesizing system. Thus, we do not know whether a system involving, say, D-nucleotides and D-amino acids, or D-nucleotides and both D- and L-amino acids, could have evolved as easily and functioned as well as the contemporary system.

(4) By using one set of enantiomers rather than two, it may be possible to achieve an important economy in the number of enzymes required. Thus the conversion of D-ribose to D-deoxyribose requires one enzyme, but the conversion of D-ribose to D-deoxyribose and L-ribose to L-deoxyribose would certainly require two. The conversion of D-ribose to L-deoxyribose would involve several enzymes.

Even if we accept that organisms with "all-D" or "all-L" proteins are at a selective advantage over "mixed" organisms, we must still ask why "all-D" organisms do not occur. An "all-D" organism would certainly function as well as an "all-L" organism. There are two possibilities. We know that if life arose only once, it must have been either D or L. We can explain the absence of all D organisms on the earth if we suppose that, by chance, the first organism was all L. The second alternative is that life arose many times on the earth and that both D and L organisms evolved on several occasions. In that case an all L organism must have developed a sufficient advantage to eliminate all competitors. The ability to synthesize an extra amino acid or vitamin, for example, might have provided such an advantage.

The origin and survival of D organisms or L organisms on a planet are equally probable. It follows that if life similar to ours exists on other planets, organisms with D-amino acids should occur on about half, and organisms with L-amino acids on the others. The same would apply to the nucleic acids. It is just possible that "all D" and "all L" organisms could coexist on some planets.

References

R. Bentley, *Molecular Asymmetry in Biology*, Volumes 1 and 2 (Academic Press, New York, 1969 and 1970.) An exhaustive discussion of optical activity.

G. H. Wheland, *Advanced Organic Chemistry*, 2nd ed. (John Wiley, New York, 1949), pp. 130–328.

P. D. Ritchie, "Recent Views on Asymmetric Synthesis and Related Processes," *Adv. Enzymology* **7**, 65 (1947). Discussions of optically active compounds and asymmetric syntheses.

R. M. Secor, *Chem. Rev.* **63**, 297 (1963). A review on the resolution of optical isomers by crystallization procedures.

W. Kuhn and E. Knopf, *Naturwissenschaften* **18**, 183 (1930); *Z. Physik. Chem.* **B7**, 292 (1930). The first successful asymmetric synthesis using circularly polarized light.

C. Frondel, *Dana's System of Minerology*, Vol. 3 (John Wiley, New York, 1962), p. 17. The abundance of quartz crystals is discussed.

G. Wald, "The Origin of Optical Activity," *Ann. N.Y. Acad. Sci.* **69**, 352 (1957). The idea that the origin of optically active compounds in biology occurred at about the same time as the origin of life rather than by prior asymmetric synthesis is presented clearly for the first time.

J. L. Bada, *J. Am. Chem. Soc.* **95**, 1371 (1972); *Adv. Chem. Ser.* **106**, 309 (1971). The rates of racemization of several amino acids are given.

H. Schneider-Bernloehr, R. Lohrmann, L. E. Orgel, J. Sulston, and B. J. Weimann, "Partial resolution of DL-adenosine by template synthesis," *Science* **162**, 809 (1968).

Chapter Fourteen

Biochemical Evolution

INTRODUCTION

The very first organisms must presumably have grown and reproduced at the expense of preformed organic materials in the prebiotic soup. This seems very obvious now, but until Oparin's book appeared, many writers believed that the first organism arose by a very improbable event and was from the very beginning capable of synthesizing all its components from CO_2 and water by photosynthesis. Sooner or later, the demand for essential components of the prebiotic soup must have come to exceed the supply. When this happened the further expansion of life depended on the ability of organisms to find ways of synthesizing their own constituents. They may have used as starting materials simple organic compounds that remained abundant in their aqueous environment (primitive metabolism) or components of the atmosphere (primitive photosynthesis). Modern biosynthetic pathways have presumably developed by the elaboration of these primitive processes.

In a similar way, the earliest living things must have used energy-rich organic molecules formed abiotically in their environment to accomplish protein synthesis and nucleic-acid replication. Here also the supply of suitable ready-made molecules must soon have proved inadequate to support the rapid expansion of life. At this point, the only convenient external source of energy, radiation from the sun, must have been harnessed in a process analogous to photosynthesis or photophosphorylation.

No fossil record of these early stages in the evolution of biochemistry has survived, so we must use indirect arguments to infer what happened on the primitive earth. A number of strategies have been used to attack this problem, but none of them has led to firm conclusions. The discussion that follows is, therefore, tentative and is meant only to illustrate some of the possibilities and to describe some of the methods which may eventually enable us to choose between them.

PRIMITIVE ORGANISMS AND PRIMITIVE BIOCHEMISTRY

If we could recognize some class of organisms as the most primitive among those now living, we would perhaps be justified in concluding that their metabolism reflects the structure of biochemistry as it first evolved on the earth. This idea has attracted a great amount of attention, but in practice it has not proved easy to apply. The difficulty lies in the problem of recognizing "primitiveness."

It seems quite certain that procaryotes, bacteria and blue-green algae, are more closely related to the first inhabitants of the earth than are the more complicated eucaryotic organisms. However, the procaryotes form a very diverse group, among which almost all known metabolic and biosynthetic pathways can be recognized. There seems no way of deciding, from the external structure of an organism or from the overall simplicity or complexity of its metabolism, whether or not the organism is primitive.

It has often been argued that since the early atmosphere was reducing, the first bacteria must have been anaerobic. It is then inferred that, among modern bacteria, the anaerobic species must be the most primitive. This argument is no longer thought to be very powerful by experts in bacterial taxonomy. They agree that modern anaerobic species may have descended almost unchanged from primitive bacteria through a series of anaerobic organisms, but they do not exclude the possibility that anaerobic species are descendants of aerobic species that have recently readapted to anaerobic environments.

Arguments based on the supposed primitiveness of particular reaction sequences impress us as being more persuasive than the arguments cited above. The classical example concerns the derivation of energy by the breakdown of glucose. The most efficient way of utilizing glucose is by oxidative phosphorylation; for each molecule of glucose oxidized to carbon dioxide, 36 molecules of ATP are formed. On the other hand only two molecules of ATP are produced during the nonoxidative, glycolytic conversion of glucose to lactic acid or ethyl alcohol and CO_2. Despite the inefficiency of glycolysis, compared to oxidative phosphorylation, glycolysis is widely distributed among living organisms, suggesting that it is a very primitive mechanism.

These facts have a plausible explanation when viewed in the light of the supposed early history of the earth's atmosphere and oceans. We have argued that at first the earth's atmosphere was reducing, and that the oceans contained sugars formed by the self-condensation of formaldehyde. Pentoses, particularly ribose, must have been used up for the synthesis of nucleic acids, but glucose and the other hexoses were probably available. Under the circumstances, one of the simplest methods available for the generation of energy was the nonoxidative metabolism of the hexoses. Primitive, and probably inefficient, processes of this kind finally evolved into the glycolytic pathway.

The reverse of this argument was used by Haldane to postulate that the early atmosphere of the earth was nonoxidizing (he proposed an atmosphere of CO_2 and NH_3). Haldane noticed that the glycolytic pathway from glucose to lactic acid was anaerobic and only the steps beyond lactic acid involved the use of O_2. On the assumption that the first steps in the oxidation of glucose were the more primitive, it followed that glycolysis steps from glucose to lactic acid arose when conditions were anaerobic and that the oxidative steps beyond lactic acid arose when O_2 appeared in the atmosphere.

The search for "primitiveness" has also been carried out at the level of protein structure and protein function. To demonstrate a typical argument concerning "primitiveness," it is convenient to compare two iron-containing proteins, hemoglobin and ferredoxin.

Hemoglobin is a protein whose prosthetic group is a Fe^{++} derivative of a complex organic molecule, protoporphyrin. The function of hemoglobin is to transport oxygen to the tissues in relatively large animals. This function cannot have been important for the origins of life since the first organisms consisted of a single cell that lived in an environment free of oxygen. Clearly we can classify hemoglobin as a modern protein which can tell us little or nothing about the origins of life.

The ferredoxins are a group of low molecular-weight proteins that contain iron bound to inorganic sulfide. They function in the reversible transfer of electrons under potentials as reducing as molecular hydrogen, and are important, for example, in photosynthesis. It has been claimed that the amino acid sequence of the ferredoxins can be shown to be derived from a simple repeating tripeptide sequence (the evidence for this claim is not very strong). However, it is true that most of the complicated amino acids are absent from ferredoxin. These facts taken together make it seem reasonably likely that the ferredoxins are indeed a very ancient family of enzymes, that they evolved when the atmosphere was strongly reducing, and that in some sense they represent "solubilized iron sulfide." It is interesting that ferredoxins or closely related proteins are involved in photosynthesis and nitrogen fixation. The enzyme, hydrogenase, which interacts directly with molecular hydrogen, is also believed to be an iron sulfide enzyme.

At this point, we should draw attention to a very active area of biological research that is concerned with the evolution of protein sequences. Corresponding proteins of different organisms have related but not identical sequences. The extent of the differences between sequences gives reliable estimates of the "relatedness" of the species concerned. In fact, it is possible to work out a "family tree" for living organisms using nothing but sequence data for cytochrome C. The family tree derived in this way agrees well with that deduced from anatomical and palaeontological evidence. Work of this kind is of the utmost importance in understanding the evolution of species, but so far it has not thrown much light on the origins of life. References to reviews of this work are provided at the end of this chapter.

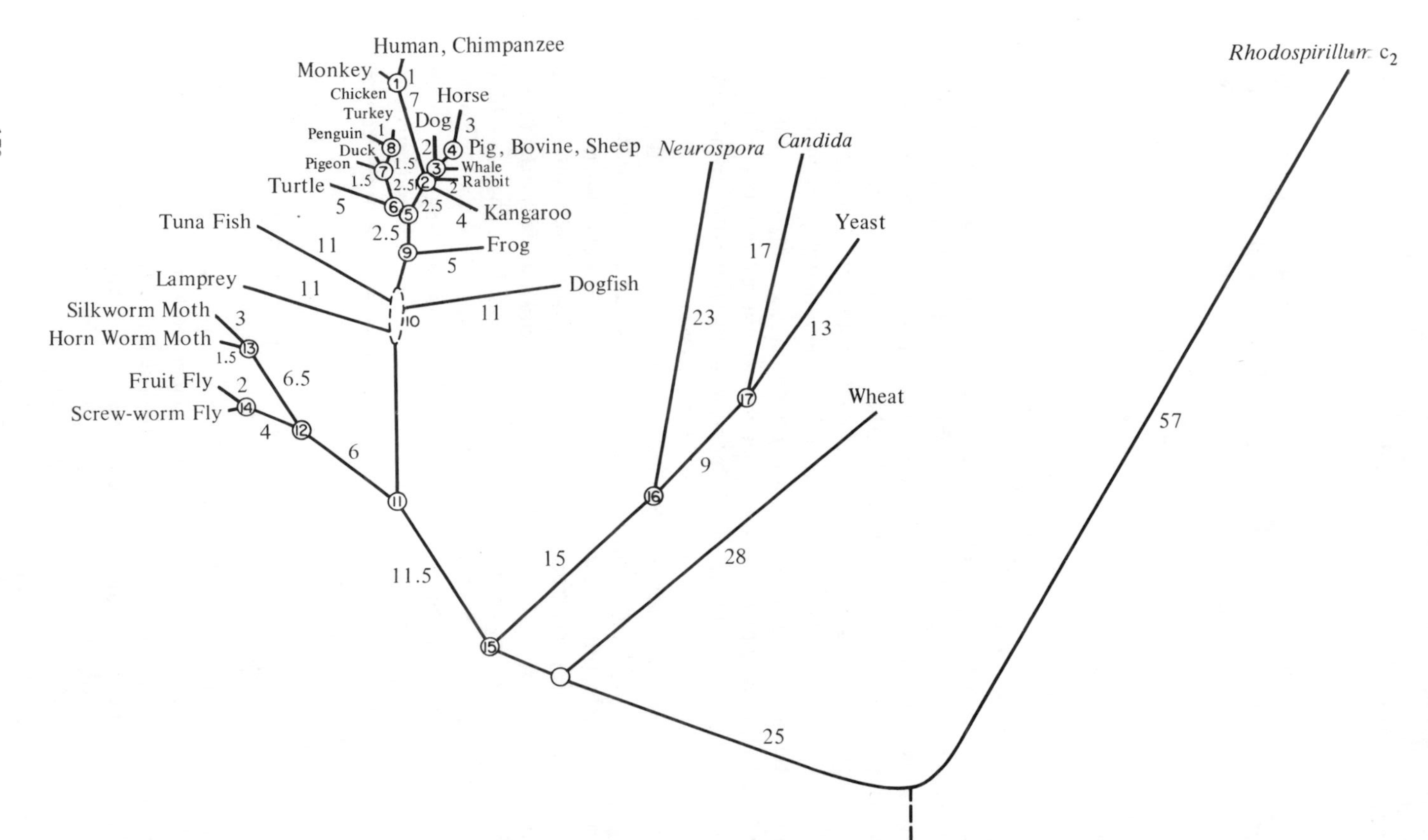

Fig. 14-1. Phylogenetic tree of cytochrome c. The numbers of inferred amino acid changes per 100 links are shown on the branches of the tree. The point of earliest time cannot be determined directly from the sequences; it has been placed by assuming that, on the average, species change at the same rate. (From M. O. Dayhoff, in *Exobiology*, C. Ponnamperuma, ed., North Holland, Amsterdam, 1972.)

At present, it must be admitted that reconstructions of the earliest phases in biochemical evolution are always highly speculative and are never unique. The next section, therefore, is to be regarded as a typical "scenario" for biochemical evolution. Many alternatives have been proposed, and several of them are every bit as convincing (or unconvincing).

A MODEL FOR BIOCHEMICAL EVOLUTION

Stage 1. The first organisms were derived from the preformed components of the prebiotic soup. A primitive form of nucleic acid replication and protein synthesis evolved before any essential component of the prebiotic soup was exhausted. Free energy was derived from prebiotic condensing agents such as cyanamide or inorganic polyphosphates.

Stage 2. A few essential components of the prebiotic soup became scarce, but organisms evolved that could supply the deficiencies by modifying other related compounds in their environment. Perhaps glycine became scarce, for example, but glycolic acid was still abundant, because it had not been incorporated into living organisms. During this stage, an organism that had "learned" to convert glycolic acid to glycine would have had an important selective advantage over its competitors. The same argument would apply to the formation of any amino acid from the corresponding hydroxy acid. Presumably the first metabolic pathways evolved at this stage in the development of biochemistry. We shall discuss this topic in more detail in the next section.

At about the same time, the first energy-producing metabolic pathways must have appeared. A primitive form of fermentation could have been achieved with enzymatic apparatus about as complicated as that required for primitive biosynthesis. The first energy-producing reaction may have

STAGE	CARBON METABOLISM	NITROGEN METABOLISM	ENERGY PRODUCTION
1	Use of prebiotic organic compounds	Use of prebiotic organic compounds and NH_3	Use of prebiotic polymerizing reagents
2	Development of a few enzymes to carry out one or two step biochemical syntheses	Use of prebiotic organic compounds and NH_3	Fermentation reactions
3	Additional biochemical steps	Reduction of N_2 to NH_3	Photophosphorylation Photoreduction with H_2S, H_2, etc.
4	Additional biochemical steps	---	Photoreduction with evolution of O_2
5	Protective mechanisms from O_2 and H_2O_2	---	Development of parts of oxidative phosphlorylation scheme
6	---	---	Completion of oxidative phosphorylation scheme
7	Development of eucaryotic organisms		Oxidative phosphorylation

Fig. 14-2. One scenario for the evolution of metabolism.

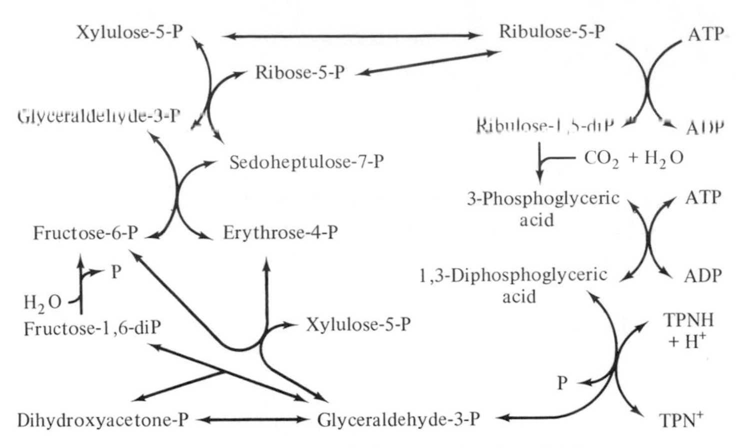

Fig. 14-3. The Calvin-Benson cycle for the reduction of CO_2 to sugars.

been glycolysis, but there are other possibilities—for example the derivation of energy from the reduction of CO_2 to methane by molecular hydrogen, a reaction that is carried out by modern methane bacteria, or the production of butyrate from ethanol and acetate (this reaction occurs in *Clostridium kluyveri*).

$$CO_2 + 4H_2 \rightarrow CH_4 + 2H_2O$$

$$CH_3{-}CH_2{-}OH + CH_3{-}COO^- \rightarrow CH_3{-}CH_2{-}CH_2{-}COO^- + H_2O$$

When organisms first began to make essential components for themselves, the evolution of a selectively permeable membrane would have been essential. Although some type of membrane may have evolved earlier, it is likely that the selectively permeable membrane came into existence at least by this stage of biochemical evolution.

Stage 3. The most widely distributed metabolic pathway for the photochemical reduction of CO_2 to the level of H_2CO is known as the *Calvin-Benson cycle*, but there are also several other photochemical pathways for fixing and reducing CO_2 in microorganisms. In the Calvin-Benson cycle, three molecules of ATP are required for the formation of one glucose molecule. Organisms also require large amounts of ATP to carry out many other metabolic reactions.

In photosynthetic organisms, the ATP needed to carry out biosynthetic reactions is obtained from photophosphorylation.

$$ADP + H_2PO_4^- \xrightarrow{h\nu} ATP + H_2O$$

It seems likely that the photochemical production of energy (photophosphorylation) developed before the photochemical production of reduced carbon compounds from carbon dioxide (photosynthesis). ATP from photophosphorylation would have permitted the transformation of simple organic compounds, which were present in large amounts in the primitive ocean

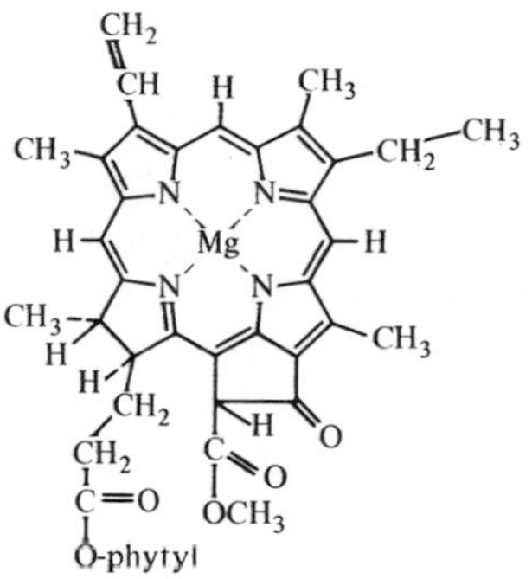

Chlorophyll a

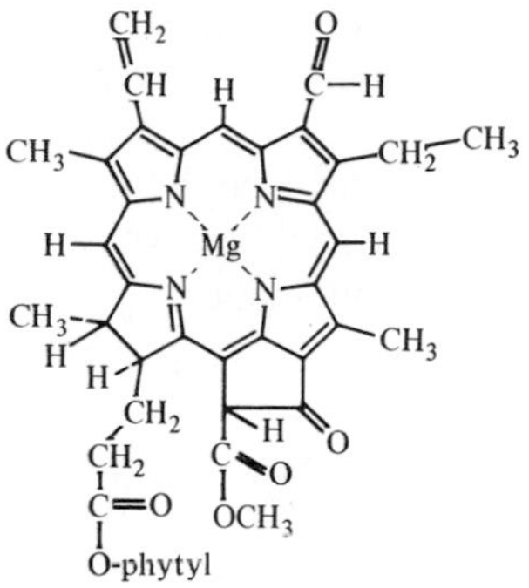

Chlorophyll b

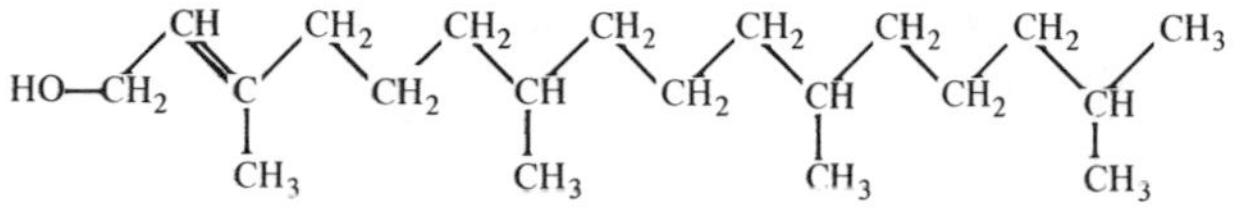

(Phytol)

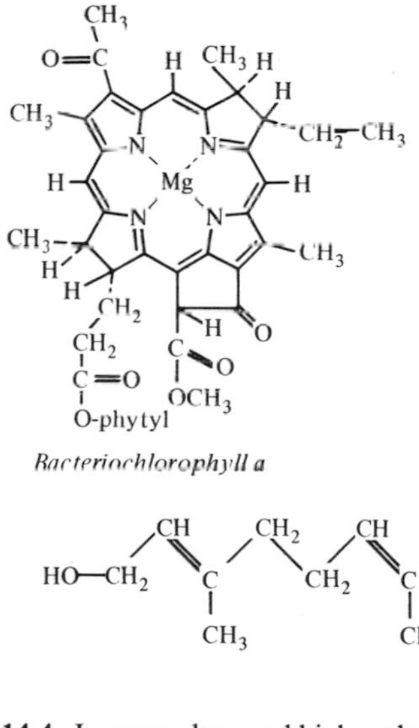

Bacteriochlorophyll a

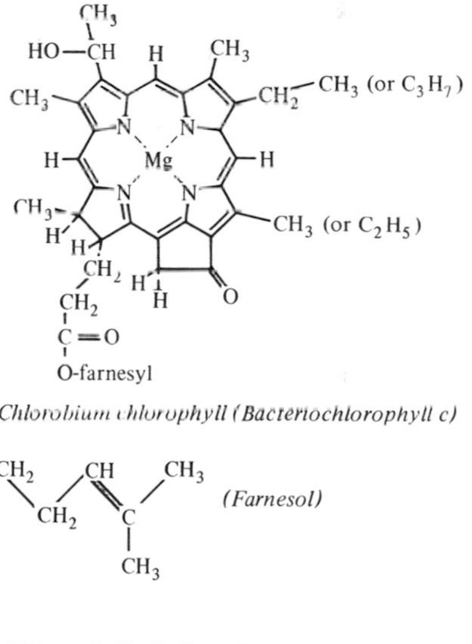

Chlorobium chlorophyll (Bacteriochlorophyll c)

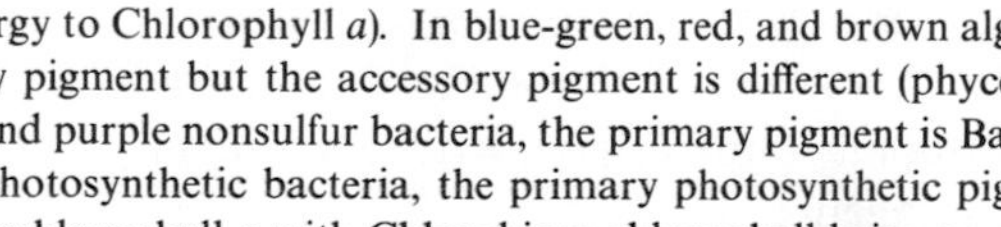

(Farnesol)

Fig. 14-4. In green algae and higher plants Chlorophyll *a* is the primary photosynthetic pigment and Chlorophyll *b* is an accessory pigment (that is, it absorbs light and transfers the energy to Chlorophyll *a*). In blue-green, red, and brown algae Chlorophyll *a* is the primary pigment but the accessory pigment is different (phycobilin). In most purple sulfur and purple nonsulfur bacteria, the primary pigment is Bacteriochlorophyll *a*. In green photosynthetic bacteria, the primary photosynthetic pigment is believed to be Bacteriochlorophyll *a* with Chlorobium chlorophyll being an accessory pigment.

(e.g., acetic acid and formic acid), into more complex organic compounds. Glucose, for example, could have been derived from acetate or formate.

$$3CH_3COOH + \text{energy} \rightarrow C_6H_{12}O_6$$

$$12HCOOH + \text{energy} \rightarrow C_6H_{12}O_6 + 6CO_2 + 6H_2O$$

Thus a great supply of organic material in the primitive ocean that was not directly useful could have been used by organisms capable of photophosphorylation, even if these organisms were not yet able to carry out photoreduction. On the other hand, an organism capable of photoreduction but not photophosphorylation would not have been able to function unless some nonphotochemical source of ATP was readily available in large amounts. Since both modern processes are very complicated, these arguments are not conclusive.

We can be rather certain that the earliest reductive photochemical processes did not split water, since this reaction requires a great amount of energy. Oxygen-evolving photosynthesis involves two photosystems because of the energy requirements, while anaerobic photosynthesis uses only one photosystem. Hydrogen sulfide was probably abundant on the surface of the primitive earth, and it seems possible that it was the first reducing agent for photosynthesis. The reaction involved is

$$CO_2 + 2H_2S \xrightarrow{h\nu} (HCHO) + 2S + H_2O$$

This is the principal photosynthetic reaction in the Green Bacteria (Chlorobacteriacae—e.g., *Chlorobium limicola*), which are strict anaerobes.

The oxidation of hydrogen sulfide to sulfate

$$2CO_2 + H_2S + 2H_2O \xrightarrow{h\nu} 2(HCHO) + SO_4^{-2} + 2H^+$$

involves more steps than oxidation to sulfur, and so this photosynthetic reaction would have developed later. The photochemical oxidation of hydrogen sulfide to sulfate takes place in the Purple Sulfur Bacteria (Thiorhodaceae—e.g., *Chromatium vinosum*).

Molecular hydrogen and organic compounds can also be used as hydrogen donors in bacterial photosynthesis.

$$CO_2 + 2H_2 \xrightarrow{h\nu} (HCHO) + H_2O$$

$$CO_2 + 2CH_3{-}\overset{\overset{\displaystyle OH}{|}}{C}H{-}CH_3 \xrightarrow{h\nu} (HCHO) + 2CH_3{-}\overset{\overset{\displaystyle O}{\|}}{C}{-}CH_3 + H_2O$$

These photochemical reactions take place in the Purple Nonsulfur Bacteria (Athiordodaceae—e.g., *Rhodospirullum rubrum*) and also in some species of Purple Sulfur Bacteria. A variety of organic reducing agents can be used by these bacteria. Ammonia and ferrous iron also seem to be plausible photosynthetic reducing agents

$$3CO_2 + 4NH_3 \xrightarrow{h\nu} 3(HCHO) + 2N_2 + 3H_2O$$

$$CO_2 + 4Fe^{+2} + 3H_2O \xrightarrow{h\nu} (HCHO) + 4Fe^{+3} + 4OH^-$$

since during this stage they would have been abundant on the earth. However, no photosynthetic bacteria that carry out these reactions have been found.

The order in which we have presented these reactions is a possible order for their development during early evolution. However, other sequences of development are also plausible.

The use of water as a reducing agent

$$CO_2 + 2H_2O \xrightarrow{h\nu} (HCHO) + O_2 + H_2O$$

is the reaction used by the prokaryotic Blue-Green Algae, the eukaryotic Green Algae, and higher plants. The development of this type of photosynthesis occurred in Stage 4.

Ammonia must have disappeared from the oceans by the time O_2 appeared in the atmosphere, and large quantities of NH_3 may have undergone photochemical destruction long before this time (Chapter 4). Thus, although the enzymatic mechanisms of nitrogen-fixation are complicated (and not worked out in any detail at the present time), this process may have been one of the earliest biochemical pathways to evolve.

If NH_3 disappeared from the oceans long before the appearance of O_2 in the atmosphere, N_2 must have first been fixed by reduction to ammonia. When conditions became more oxidizing, nitrate and nitrite may have appeared in the environment. These could be used as nitrogen sources, as soon as the pathway for their reduction developed. A possible sequence for the development of nitrogen metabolism consistent with this "scheme" is:

$$\tfrac{1}{2}N_2 + 3H \rightarrow NH_3 \qquad (1)$$

$$NO_2^- + H^+ + 6H \rightarrow NH_3 + 2H_2O \qquad (2)$$

$$NO_3^- + H^+ + 8H \rightarrow NH_3 + 3H_2O \qquad (3)$$

An alternative possibility is that the NH_3 in the ocean was not completely depleted until the time that the O_2 appeared in the atmosphere. The nitrates and nitrites would have been produced by thunderstorms, and the first metabolic pathways used by bacteria to obtain nitrogen would have been the reduction of nitrites or nitrates to ammonia. In that case, reactions (2) and (3) would have developed before reaction (1).

It is worth pointing out that the reduction of nitrate is chemically less difficult than the reduction of N_2, and perhaps was also less difficult enzymatically. However, since the amount of nitrate and nitrite produced by thunderstorms is rather limited, the number of living organisms that could have been supported by nitrate formed in thunderstorms would have been correspondingly limited.

Stage 4. When all abundant reducing agents other than water had been used up, oxygen-evolving photosynthesis would, of necessity, have developed. This led gradually to the accumulation of O_2 in the atmosphere and consequently to a complete change in the chemistry of the primitive atmosphere

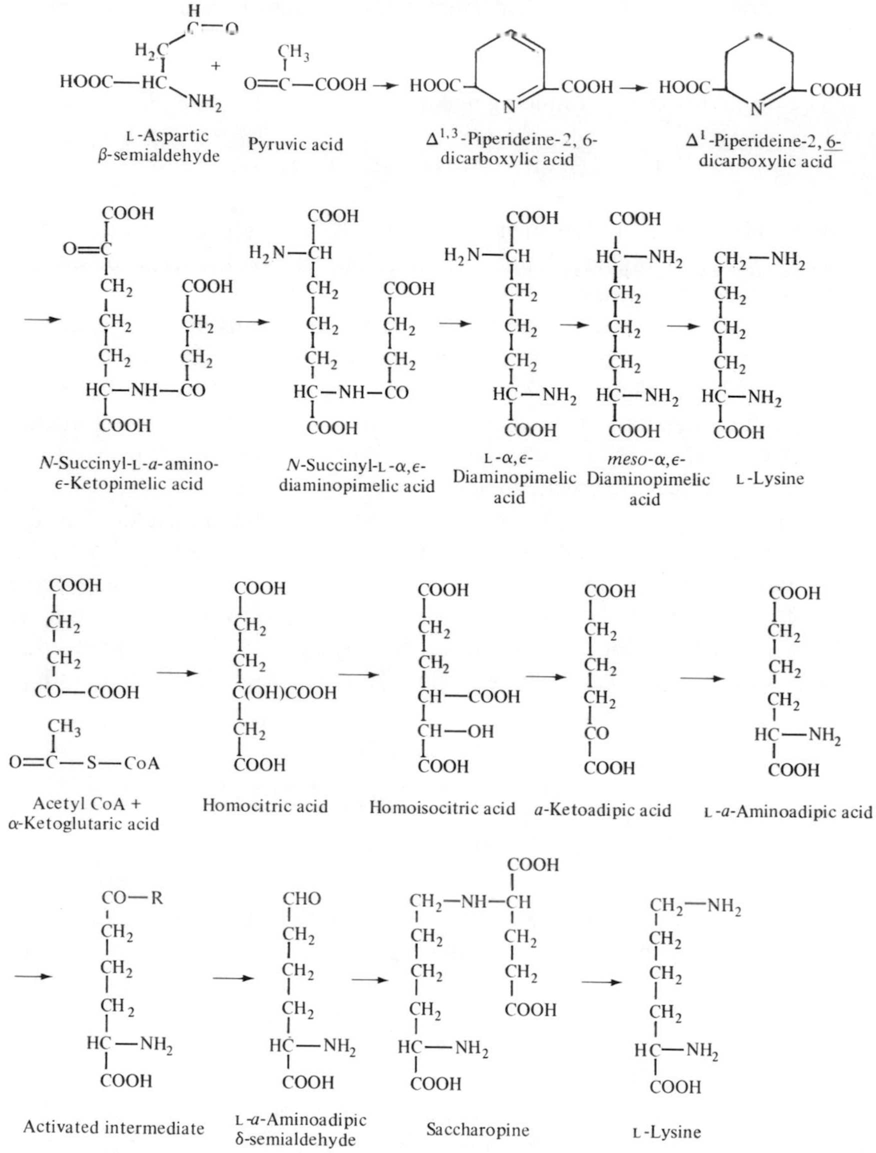

Fig. 14-5. The two pathways of lysine biosynthesis. (Upper) The diaminopimelic acid pathway. (Lower) The aminoadipic acid pathway.

and oceans. Prebiotic synthesis was no longer possible (it may have terminated much before this) and the survival of life became dependent on oxygen-evolving photosynthesis (except in isolated areas where reducing conditions prevailed).

Stage 5. The abundance of oxygen in the atmosphere increased until it began to interfere with normal metabolism, which had, of course, evolved under reducing conditions. The flavin coenzymes and sulhydryl groups are particularly sensitive to molecular oxygen. The product of these non-enzymatic reactions is frequently H_2O_2, which destroys many of the organic constituents of living organisms. It has been suggested that a special protective system, the peroxysome, evolved at this stage to prevent organisms from dying through oxygen-poisoning.

Stage 6. The oxygen abundance evolved further until it became adequate to support a new type of energy production, oxidative phosphorylation. The oxidation of glucose by oxygen to carbon dioxide and water was used to recover much of the energy stored during photosynthesis. Most living forms soon became completely dependent on oxidative-phosphorylation—anaerobic forms survived only in special environments.

Stage 7. Once an efficient oxygen-utilizing form of energy metabolism got started, eucaryotic organisms and multicellular organisms evolved.

THE EVOLUTION OF INTERMEDIARY METABOLISM

In a few cases, biosynthetic pathways differ in different groups of organisms. Lysine synthesis, for example, follows the diaminopimelate route in bacteria, some fungi and plants, and the α-aminoadipate route in other fungi and euglenids. Animals do not synthesize lysine. In the majority of cases, however, the same biosynthetic pathways occur in all organisms. This is true, for example, for the synthesis of aromatic amino acids, purine and pyrimidine nucleosides, and saturated long-chain fatty acids. It seems probable, therefore, that these pathways achieved their final form early in the history of life. How did such pathways evolve?

It has often been suggested that metabolic pathways parallel the corresponding prebiotic syntheses that occurred on the primitive earth. It is not difficult to show that this hypothesis cannot be correct in the majority of cases. Perhaps the strongest evidence comes from a direct comparison of known contemporary biosynthetic pathways with reasonable prebiotic pathways—in general, they do not correspond at all. This is perhaps not too surprising since the compounds that are readily formed in the atmosphere in prebiotic reactions are not always made readily by enzymatic reactions. Cyanoacetylene and phenylacetylene are two molecules that are readily formed in the gas phase, but which could only be obtained with difficulty in living organisms.

It seems likely that the later steps in a biosynthetic pathway evolved before the first steps. Each enzyme, as it evolved, must have conferred a selective advantage to the organism in which it evolved. Biosynthetic pathways from simple starting materials to final products, such as amino acids,

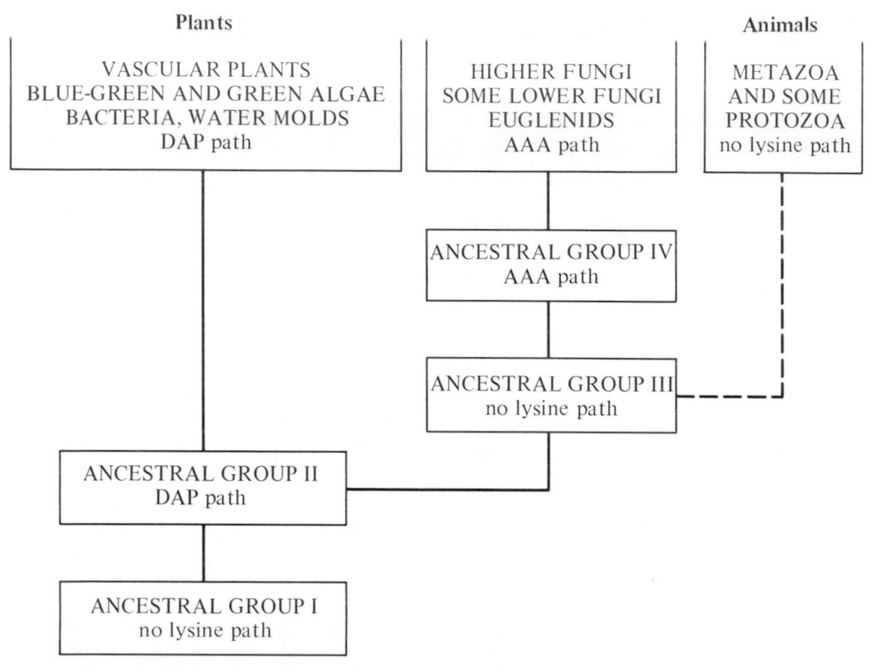

Fig. 14-6. The evolutionary distribution of lysine pathways. H. J. Vogel concluded that the evolutionary descent of the paths of lysine synthesis involves four ancestral groups. The earliest organisms, Group I, depended on an external supply of lysine. Group II developed the first lysine path using α,ε-diaminopimelic acid (DAP). Group III underwent an evolutionary loss of lysine synthesis, presumably because lysine was then more readily available. Group IV contains ancestral forms that gained an advantage under their particular evolutionary conditions in being able to form lysine (this time via α-aminoadipic acid—AAA). In the more recent precursors of modern animals, the ability to synthesize lysine apparently remained lost.

involve many steps. The evolution of an enzyme that carried out an early step in such a sequence would be useless until the later enzymes were available because there would be no use for the product of such an enzyme. We must conclude, therefore, that in most cases the final steps in metabolic sequences evolved before the earlier ones.

A detailed discussion of this problem was first given by Horowitz. He supposed that the prebiotic soup contained compounds that were directly useful to primitive organisms, for example, the amino acids and bases, and also a variety of related compounds which, while not useful in themselves, could easily be converted to useful biochemicals. Once an essential biochemical became depleted, pathways evolved to produce the needed compound from some closely related substance in the prebiotic soup. This hypothesis is consistent in some but not all cases with what we know about biosynthetic pathways.

The prebiotic reaction of HCN and NH_3 which produces adenine gives in addition 4-amino-imidazole-5-carboxamide. Thus the prebiotic soup is likely to have contained this substance as a potential precursor of adenine when the supply of adenine, itself, was used up. It is interesting that 4-amino-imidazole-carboxamide (as the ribotide) is the immediate precursor of the purines in contemporary organisms.

The further steps in the purine synthesis pathway can be considered (written as free compounds rather than in the form of ribotides)

glycine $\xrightarrow{1}$ glycine amide $\xrightarrow{2}$ N-formyl glycine amide $\xrightarrow{3}$

N-formamido glycine amide $\xrightarrow{4}$ aminoimidazole $\xrightarrow{5}$

aminoimidazole carboxylic acid $\xrightarrow{6}$ aminoimidazole carboxamide $\xrightarrow{7}$

formamido aminoimidazole carboxamide $\xrightarrow{8}$ hypoxanthine $\xrightarrow{9}$ adenine.

If aminoimidazole carboxamide was present in the prebiotic environment, then aminoimidazole carboxylic acid would also have been present, since it is a hydrolysis product. Therefore, an enzyme to carry out reactions 6 and 7 would have been developed after the enzymes carrying out reactions 8 and 9. Aminoimidazole is also likely to have been in the primitive ocean since it is in equilibrium with the carboxylic acid in the presence of bicarbonate ion. N-Formamidoglycine amide does not seem particularly plausible as a constituent of the primitive ocean, since amidines hydrolyze rather rapidly, but small quantities might have been made under dehydrating conditions. N-Formyl glycine amide, glycine amide, and glycine are all very plausible constituents of the primitive ocean.

It seems likely that the further a precursor is from adenine in this pathway, the greater its concentration (with a few exceptions) would have been in the primitive ocean. Thus the enzymes of the purine biosynthetic pathway would have developed from the last step in sequence to the first step.

A slight extension of Horowitz's theory suggests that intermediary metabolism may, in many cases, reverse the pathway by which important biochemicals were degraded in the prebiotic soup. In general, the first step in the decomposition of a complex biochemical will lead to a product with a structure closely related to the original biochemical. In such cases, a single chemical step will often be adequate to reverse the spontaneous decomposition and so to recover the original biochemical. Enzymes could easily have developed to bring about simple conversions of this kind.

In time, the first decomposition product would have been used up, but then a second decomposition product would often have been available. Since the pathway from the first decomposition product would already have evolved, only one new enzyme would have been needed to convert the second decomposition product all the way to the desired product. Repetition of this process could have led to the step-by-step elaboration of complex metabolic pathways. The pathway of purine synthesis does follow fairly closely the pathways expected for the hydrolysis of purines under alkaline conditions.

In other cases, no obvious relationship exists between simple hydrolytic pathways and biosynthetic pathways. Perhaps in some of these cases the synthetic pathway parallels pathways of decomposition that involved components of the prebiotic soup. The biosynthesis of pyrimidines, for example, may parallel their photochemical decomposition in the presence of hydrogen cyanide. A somewhat similar and equally speculative scheme for the evolution of the pathway to the aromatic amino acids can be written.

We may conclude, then, that biosynthetic pathways developed to carry out chemical reactions that occur relatively easily in the absence of an enzyme. Sometimes a compound related to a required biochemical was available as a result of a parallel synthesis in the prebiotic soup. On other occasions, a decomposition product of the biochemical was available. In either case, an enzyme could easily evolve that would bring about the required synthesis in a single step without the simultaneous development of other enzymes. No doubt other mechanisms were also important. If the prebiotic pathway included an inefficient, rate-limiting step, then an enzyme that catalyzed that step would have been useful and would have been selected for. Occasionally, this must have led to biosynthetic pathways that paralleled prebiotic pathways, but as we have seen such cases seem to have been rare.

References

Prokaryotic and eukaryotic organisms

Organization and Control in Prokaryotic and Eukaryotic Cells, *Symp. Soc. Gen. Microbiol.* **20** (Cambridge University Press, Cambridge, 1970).

R. Y. Stanier, "Some aspects of the biology of cells and their possible evolutionary significance," *Symp. Soc. Gen. Microbiol.* **20**, 1–38 (1970).

L. Sagan, "On the origin of mitosing cells," *J. Theor. Biol.* **14**, 225 (1967).

L. Margulis (Sagan), *Origin of Eukaryotic Cells* (Yale University Press, New Haven, Conn., 1970).

P. H. Raven, "A multiple origin for plastids and mitochondria," *Science* **169**, 641 (1970).

Protein sequences, ferredoxins, nitrogen fixation

M. O. Dayhoff, *Atlas of Protein Sequence and Structure*, Vol. 4 (National Biomedical Foundation, Silver Springs, Maryland, 1969).

A. San Pietro, ed., *Non-Heme Iron Proteins* (Antioch Press, Yellow Springs, Ohio, 1965).

B. B. Buchanan and D. I. Arnon, "Ferredoxins: Chemistry and function in photosynthesis, nitrogen fixation, and fermentative metabolism," *Adv. Enzymol.* **33**, 119–176 (1970).

D. O. Hall, R. Cammack, and K. K. Rao, "Role for ferredoxins in the origin of life and biological evolution," *Nature* **233**, 136 (1971).

J. R. Benemann and R. C. Valentine, "The pathways of nitrogen fixation," *Adv. Microbiol. Physiol.* **8**, 59–104 (1972).

Photosynthesis

C. B. van Niel, "The bacterial photosyntheses and their importance for the general problem of photosynthesis," *Adv. Enzymol.* **1**, 263 (1941).

C. B. van Niel, "The comparative biochemistry of photosynthesis," in *Photosynthesis in Plants*, J. Franck and W. E. Loomis, eds. (The Iowa State College Press, Ames, Iowa, 1949), pp. 437–495.

M. Calvin and J. A. Bassham, *The Photosynthesis of Carbon Compounds* (Benjamin Press, New York, 1962).

D. I. Arnon *et al.*, "Photoproduction of hydrogen, photofixation of nitrogen and a unified concept of photosynthesis," *Nature* **190**, 601 (1961).

M. Evans and F. Whatley, "Photosynthetic mechanisms in prokaryotes and eukaryotes," *Symp. Soc. Gen. Microbiol.* **20**, 203–220 (1970).

S. Granick, "Speculations on the origins and evolution of photosynthesis," *Ann. N.Y. Acad. Sci.* **69**, 292 (1957).

J. M. Olsen, "The evolution of photosynthesis," *Science* **168**, 438 (1970).

H. Gaffron, "The role of light in evolution: The transition from a one quantum to a two quantum mechanism," in *The Origin of Prebiological Systems*, S. W. Fox, ed. (Academic Press, New York, 1965), pp. 437–460.

Peroxisome

C. de Duve, "The peroxisome, a new cytoplasmic organelle," *Proc. Roy. Soc.* **B173**, 71 (1969).

Lysine metabolism, energy metabolism, primitive metabolism

H. J. Vogel and R. H. Vogel, "Some glimpses of evolution," *Chem. Eng. News*, 90–97 (Dec. 11, 1967).

H. J. Vogel, J. S. Thompson, and G. D. Shockman, "Characteristic metabolic patterns of prokaryotes and eukaryotes," *Symp. Soc. Gen. Microbiol.* **20**, 107–119 (1970).

E. Broda, "The evolution of bioenergetic processes," *Prog. Biophys. Mol. Biol.* **21**, 143 (1970).

H. Baltscheffsky, "Inorganic pyrophosphate and the origin and evolution of biological energy transformation," in *Chemical Evolution and the Origin of Life*, R. Buvet and C. Ponnamperuma, eds. (North Holland, Amsterdam, 1971), pp. 466–474.

F. Lipmann, "Projecting backwards from the present stage of evolution of biosynthesis," in *The Origins of Prebiological Systems*, S. W. Fox, ed. (Academic Press, New York, 1965), pp. 259–280.

E. Beerstecher, Jr., "The biochemical basis of chemical needs," in *Comparative Biochemistry*, M. Florkin and H. S. Mason, eds. (Academic Press, New York, 1964), Vol. VI, pp. 119–220.

N. H. Horowitz, "On the evolution of biochemical synthesis," *Proc. Nat. Acad. Sci. U.S.* **31**, 153 (1945).

N. H. Horowitz, "The evolution of biochemical synthesis—retrospect and prospect," in *Evolving Genes and Proteins*, V. Bryson and H. J. Vogel, eds. (Academic Press, New York, 1965) pp. 15–23.

C. Degani and M. Halmann, "Alkaline reactions of glucose-6-phosphate," *J. Am. Chem. Soc.* **90**, 1313 (1968).

C. Degani and M. Halmann, "Solvolysis of phosphoric acid esters. Hydrolysis of glucose-6-phosphate. Kinetic and tracer studies," *J. Am. Chem. Soc.* **88**, 4075 (1966).

C. A. Bunton and H. Chaimovich, "The hydrolysis of glucose-6-phosphate," *J. Am. Chem. Soc.* **88**, 4082 (1966).

C. D. Gutsche *et al.*, "Base catalyzed triose condensations," *J. Am. Chem. Soc.* **89**, 1235 (1967).

Chapter Fifteen

Meteorites, Comets, Interstellar Molecules, and Life on Other Planets

METEORITES

Meteorites are objects from space that fall on the earth. An object that falls through the atmosphere, but burns up entirely, is called a *meteor*. Meteors are, of course, detected because they emit light as they burn up.

There are three general classes of meteorites—the irons, the stony-irons, and the stones. Of those meteorites that are seen to fall and are then picked up, about 5% are irons, 1% are stony-irons, and 94% are stones. Of those meteorites that are not seen to fall but are found, sometimes many years after they have fallen, about 66% are irons, 7% are stony-irons, and 27% are stones. The difference in distribution is easy to understand—a stony meteorite looks much like an ordinary earth rock, but a piece of metallic iron is very noticeable. Thus stony meteorites are much less likely to be recognized.

The irons consist largely of an iron-nickel alloy containing 4–16% nickel. The irons are usually larger than the stones and they are the meteorites usually on prominent display in museums. The stony-irons are, as the name implies, partly stone and partly iron. The stones can be divided into a number of classes—the chondrites, achondrites, and carbonaceous chondrites. There are many subdivisions of these classes, but we will not be concerned with them here.

The chondrites are named for the chondrules which they contain. Chondrules are roughly spherical beads of about 1 mm diameter with a characteristic texture. They contain the minerals olivine and orthopyroxene. The chondrules may have been formed by the rapid cooling of molten particles. The achondrites are similar to the chondrites except for the absence of chondrules.

(a)

(b)

Fig. 15-1. (a) A piece of the Murchison meteorite. (b) The three classes of meteorites. On the left is an iron meteorite that has been cut in two, polished, and then etched to show the pattern of crystals of iron-nickel alloy. In the center is a stony meteorite with a grainy texture similar to a terrestrial rock. On the right is a stony-iron, its network of nickel-iron is filled with stone rich in olivine. ((a) From NASA. (b) From Life Nature Library, *The Earth*, p. 31.)

Meteorites are generally believed to have come from the asteroid belt. The iron meteorites presumably were formed in a planet large enough to have a molten core. This planet subsequently disintegrated, presumably by collision with another planet or asteroid. However, there are many other theories of meteorite formation, and it is possible or even likely that different types of meteorites come from different places.

The carbonaceous chondrites form a small but intriguing subclass of the chondrites. About 2% of all known meteorites are carbonaceous chondrites. However, carbonaceous chondrites, like other stony meteorites, are hard to recognize, and furthermore disintegrate rapidly if they are not collected. They clearly represent more than 2% of the meteorites reaching the surface of the earth. Furthermore, it seems likely that a high proportion of carbonaceous chondrites are broken up completely in the atmosphere since they are so fragile.

On falling through the atmosphere, a meteorite produces a great deal of heat and light. The surface of the meteorite is frequently melted, forming a fusion crust, but the interior of the meteorite is only slightly warmed. This is of particular concern with the carbonaceous chondrites because heating of their interiors would pyrolyze or burn up the organic compounds.

The carbonaceous chondrites are usually divided into three different types, based on their carbon content and mineralogy. Type I consists of largely amorphous hydrated silicates, sulfur as magnesium sulfate, about 20% water, and 3 to 5% organic carbon. Type II consists largely of hydrated magnesium-iron silicate, considerable elemental sulfur, 10 to 15% water, and 2 to 4% organic carbon. Type III is largely olivine, about 2% water, and 0.1 to 2% carbon, much of which may be amorphous carbon. The first attempt to analyze the organic compounds in such a meteorite was reported by Berzelius in 1834. He was not able to identify particular organic compounds in the Alais meteorite (Type I), but decided that the organic compounds were not of biological origin.

Subsequent studies during the nineteenth century and the first half of the twentieth century showed that most of the carbon in the carbonaceous chondrites is in the form of an insoluble polymer. About 1% of the total carbon is present as a complex mixture of soluble, high-molecular-weight substances. In 1961, a new study of the organic constituents of a carbonaceous chondrite, the Orgeuil meteorite (Type I) which fell in 1864, was reported. This time modern analytical techniques were used, and a variety of saturated hydrocarbons were identified by mass spectrometry. The authors concluded that their results provided evidence of biogenic activity.

This conclusion did not go unopposed and a considerable controversy developed. Some writers believed that all the compounds found on the meteorite were indigenous, but that they had been formed abiogenically. Others raised the possibility of terrestrial contamination. An even more intense controversy followed the report of "organized elements," resembling bacteria, in the meteorite. Some authors believed that they were extraterrestrial fossils, but others claimed that they were terrestrial contaminants. Alternatively, it was argued that the "organized elements" were inorganic artifacts rather than extraterrestrial microorganisms. We do not believe that there is any evidence for indigenous biogenic compounds or structures in the Orgueil meteorite. Indigenous organic compounds are present but their mode of synthesis is unclear.

A carbonaceous chondrite (Type II) fell near Murchison, Australia on September 28, 1969, and portions were collected that day. More material was

Table 15-1. Amino Acids in the Murchison Meteorite

Type of Amino Acid	μg/g of meteorite	%D Isomer[a]
Glycine	6	—
Alanine	3	50
α-Amino-n-butyric acid	~1	~50
α-Aminoisobutyric acid	~2	—
Valine	2	40–47
Norvaline	~2	~50
Isovaline	~0.3	?
Proline	1	40–43
Pipecolic acid	~0.1	~50
Aspartic acid	~2	~50
Glutamic acid	3	45
β-Alanine	~0.5	—
β-Amino-n-butyric acid	~0.1	?
β-Aminoisobutyric acid	~0.1	~50
γ-Aminobutyric acid	~0.1	—
Sarcosine	~1	—
N-Ethylglycine	~1	—
N-Methylalanine	~1	?

[a] The quantitative values marked ~ are very approximate. The amino acids with — in the %D Isomer column have no asymmetric carbon; the D and L isomers of the amino acids with ? in this column did not separate on the gas chromatography columns used in the analysis.

found during February and March of 1970. The first reports on the organic constituents of this meteorite were very exciting. The organic matter in the meteorite was hydrolyzed with hydrochloric acid and then subjected to analysis for amino acids. The amino acids were desalted by ion exchange chromatography and analyzed by gas chromatography, the most abundant of them being identified in a mass spectrometer. The first report identified glycine, alanine, sarcosine, α-aminoisobutyric acid (2-methylalanine), valine, proline, and glutamic acid. A second report identified eleven more amino acids. These are given in Table 15-1.

There are two good reasons for believing that these amino acids cannot be terrestrial contaminants. Firstly, the amino acids with asymmetric carbons are present as racemic mixtures of the D and L forms. Terrestrial contaminants are predominantly in the L form, and since the meteorite was collected promptly and handled carefully there was no opportunity for racemization to occur. Secondly, α-aminoisobutyric acid, N-ethylglycine, N-methylalanine, isovaline (2-methyl-α-amino-n-butyric acid), β-amino-n-butyric acid, and norvaline do not occur (or are rare) in biological systems. Sarcosine,

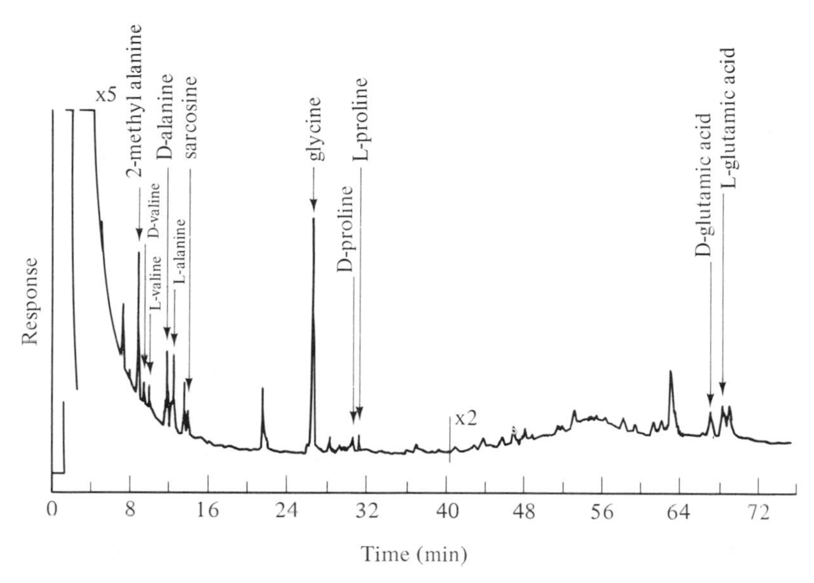

Fig. 15-2. Gas chromatogram of N-trifluoroacetyl-D-sec-butyl esters of amino acids in acid hydrolyzed water extract of a 10-gram sample of the Murchison meteorite. 2-Methyl alanine is α-aminoisobutyric acid. The peak just following L-glutamic acid is an unknown. 2-Methylalanine, sarcosine, and glycine do not have optical isomers, so only one peak can appear. Gas chromatography. 0.02 in. × 150 ft capillary column coated with UCON 75-H-90,000; temperature programmed from 100 to 150°C at 1°C/min. (From *Nature* **228**, 923, 1970.)

α-amino-n-butyric acid, pipecolic acid, β-aminoisobutyric acid, and γ-aminobutyric acid do occur in biological systems but they are not abundant. The indigenous character of the organic compounds is also established by the C^{13}/C^{12} isotope ratio; analysis of several different extracts showed that the total organic carbon in the meteorite was higher in C^{13} than any carbon that occurs on the earth. The C^{13}/C^{12} ratio, however, was not measured for the amino acids because of the small amounts present, so the C^{13}/C^{12} isotope data do not establish the indigenous character of the amino acids. However, the other evidence seems conclusive.

The amino acids found in the meteorite are strikingly similar to those formed by electric discharges acting on mixtures of methane, nitrogen, ammonia, and water (Chapter 7). Not only are the identified amino acids similar, but also the relative abundances are very similar.

After the results for the Murchison meteorite were reported, other carbonaceous chondrites, some of which had been standing on museum shelves for many years, were analyzed by gas chromatography-mass spectrometry. It was possible to show that the amino acids which had been previously detected were in part contaminants and in part indigenous. In this work it was assumed that any excess of the L isomer of an amino acid was a contaminant, and that the D isomer and an equal amount of the L isomer was indigenous.

A search for nucleic acid purines and pyrimidines in the Murchison meteorite was made, but none could be found. However, 4-hydroxypyrimidine (6 $\mu g\ g^{-1}$), 4-hydroxymethylpyrimidine (7 $\mu g\ g^{-1}$), 4-hydroxy-1 (or 6)-methylpyrimidine (3 $\mu g\ g^{-1}$) were found. Aliphatic and aromatic hydrocarbons were also found in Murchison and the evidence indicates that they are indigenous.

It is hard to assess the significance of these findings since very little is known about the origin of meteorites generally, or of the carbonaceous chondrites in particular. The facts however seem clear. The Murchison meteorite does contain large quantities of amino acids, a total of 2×10^{-7} moles of amino acids per gram. This is more than many desert sands. There is every reason to believe that they are of extraterrestrial origin. The amino acids in the meteorites correlate strikingly with those formed in prebiotic syntheses of the type discussed in Chapter 7. This indicates that our ideas about the prebiotic origin of the amino acids are at least in part correct.

COMETS

Comets are another group of objects in outer space that contain organic molecules. The head of a typical comet is believed to be 10 to 100 km in diameter which corresponds to 5×10^{17} to 5×10^{20} g, and to consist of a mixture of "ices" of H_2O, NH_3, and CH_4. The temperature in the interior of the head of a comet is probably less than 10°K so that even the methane is present as a nonvolatile solid. When a comet approaches the sun, the surface is heated up by the sun's visible radiation and also by solar protons. The interior of the head of a comet is heated only slightly because of the low thermal conductivity of the material. This causes a small fraction of the comet to vaporize forming a luminescent tail extending for millions of kilometers.

Very prominent fluorescence emission lines in the spectrum of the tails of comets come from the molecules CN and C_2. The spectral lines of C_3, CH, CH_2, NH, NH_2, OH, OH^+, CO^+, and NH_2^+ are also observed. The prominence of CN and C_2 lines does not necessarily mean that they are particularly abundant. Other molecules may be present in greater quantity but fail to absorb or emit as strongly as C_2 and CN. Even so, it seems likely that substantial quantities of cyanide, acetylene, and larger molecules are contained in the comet head, since a number of experiments in which organic ices are subjected to high energy radiation have shown that organic compounds are produced in fair yield.

We do not believe that organic compounds from comets were vital or even particularly important for the origin of life on the earth. Comets may have collided frequently with the earth when it was condensing into a solid

Fig. 15-3. Halley's Comet appears every 76 years. This picture was taken in 1910. The comet is expected again in 1986. (From Mount Wilson and Palomar Observatories.)

planet and may have added significant amounts of carbon compounds. A comet with a mass of 10^{19} g would contribute 2 g cm^{-2} of material to the earth's surface. This material would include some organic compounds, but not all those organic compounds originally present in the comet would survive, since some material would be pyrolyzed on impact. Since the compounds in the comet head seem to be just the ones that are formed under plausible conditions in a reducing atmosphere, there seems little justification for assigning an important role in the origins of life to cometary synthesis.

THE MOON

Before the Apollo landings on the moon and the return of the samples, opinion was divided as to whether there would be any organic material in lunar rocks. Some workers believed that there would be very little but others felt that lunar rocks would contain large amounts of prebiotic compounds, and might provide experimental confirmation of our ideas about the primitive earth. It was even proposed that the Maria are giant seas of asphalt.

The analyses were in some respects disappointing. The lunar rocks contain only 10 to 70 parts per million of carbon, while the soils and breccias contain 25 to 250 parts per million. Most of this carbon comes out on heating as carbon monoxide and carbon dioxide along with about 5 ppm of methane and small hydrocarbons. There is less than 1 ppm of volatile or pyrolyzable organic compounds.

Special attention was devoted to the analysis of amino acids, for which the analytical methods are extremely sensitive. The moon samples were acid hydrolyzed (or the water wash of the moon samples was acid hydrolyzed) and the amino acids were analyzed by a very sensitive amino acid analyzer and by gas chromatography. These analyses gave amino acid abundances varying from zero to as high as 70 parts per billion (10 ppb is 10^{-10} mole of amino acids per gram). It should be noted that at these low levels, the problems of contamination from reagents and dust are enormous, and the possibility that the high results are due to contamination cannot be excluded. It is generally agreed that the amino acids are not present as such, but are produced during the extraction and hydrolysis steps of the analysis. It is not understood how this comes about.

With the benefit of hindsight it is "obvious" that no organic compounds were to be expected in the lunar rocks. The moon is continuously bombarded by the solar wind (relatively high-energy protons and other elements) which would decompose any organic matter present in the surface rocks. In fact, it is believed that the carbon found in the lunar rocks comes from the solar wind. It should be noted, however, that the solar wind does not penetrate more than a few tens of meters below the surface of the moon, so that organic material below this depth would not be decomposed.

MOLECULES IN INTERSTELLAR SPACE

The study of molecules in interstellar space has been made possible by recent developments in radio astronomy. Absorption or emission spectra of molecules in space could, in principle, be examined in any region of the electromagnetic spectrum. In practice, most of the identifications so far achieved are based on microwave measurements. Work on the infrared spectra of molecules in space is just beginning. This is likely to be an exciting field for the next few years.

The molecules identified in space at the present time are listed in Table 15-2. The presence of ammonia and water calls for little comment, since under most circumstances they are expected to be the most abundant molecules containing nitrogen and oxygen, respectively. The striking feature of the list is the presence of so many of the proposed "prebiotic" precursors. The first four organic molecules found in space included three of the most discussed prebiotic precursors.

Hydrogen cyanide is involved in most proposed prebiotic syntheses of amino acids and purines. It is also one of the most abundant organic

Table 15-2. Interstellar Molecules. The wavelengths and megahertz refer to the line by which the molecule was discovered. Isotopic species other than the most abundant one have been seen for some of the molecules (e.g., $O^{18}H$, $C^{13}O$, CO^{18}, $HC^{13}N$, HCN^{15}, CS^{24}). The number of lines identified includes the isotopic species. The ultraviolet lines are seen in absorption. The radio lines are seen in emission, with a few exceptions. The concentrations of molecules are estimated roughly to be 10^4 molecules cm^{-3} for H_2, 1 molecule cm^{-3} for CO, and the remainder between 10^{-5} and 10^{-3} molecule cm^{-3}.

Year	Molecule	Formula	Wavelength (cm or Å)	Megahertz	Number of Lines Identified
1937		CH	4,300 Å	—	
1940	Cyanogen radical	CN	3,875 Å	—	
1941		CH^+	3,745–4,233 Å	—	
1963	Hydroxyl radical	OH	18.0	1,667.3; 1,667.5	10
1968	Ammonia	NH_3	1.25	23,694	7
1969	Water	H_2O	1.35	22,235	1
1969	Formaldehyde	$H_2C{=}O$	6.21	4,830	10
1970	Carbon monoxide	CO	0.260	115,271	3
1970	Cyanogen radical	CN	0.264	113,492	1
1970	Hydrogen	H_2	1,013–1,108 Å	—	
1970	Hydrogen cyanide	HCN	0.338	88,632	6
1970	X-ogen	?	0.336	89,190	1
1970	Cyanoacetylene	$HC{\equiv}C{-}CN$	3.30	9,098	3
1970	Methyl alcohol	CH_3OH	35.9	834	9
1970	Formic acid	HCOOH	18.3	1,639	1
1971	Carbon monosulfide	CS	0.20	146,967	7
1971	Formamide	$HC(NH_2)O$	6.5	4,619	1
1971	Silicon monoxide	SiO	0.23	130,268	1
1971	Carbonyl sulfide	OCS	0.27	109,463	4
1971	Acetonitrile	$CH_3C{\equiv}N$	0.27	110,330; 110,384	5
1971	Isocyanic acid	$HN{=}C{=}O$	0.34	87,925	3
1971	Methyl acetylene	$CH_3C{\equiv}CH$	0.35	85,457	1
1971	Acetaldehyde	CH_3CHO	28.1	1,065	5
1972	Thioformaldehyde	$H_2C{=}S$	9.5	3,139	1
1972	Hydrogen sulfide	H_2S	0.18	168,762	1
1972	Methylene imine	$H_2C{=}NH$	5.7	5,289; 5,292	1

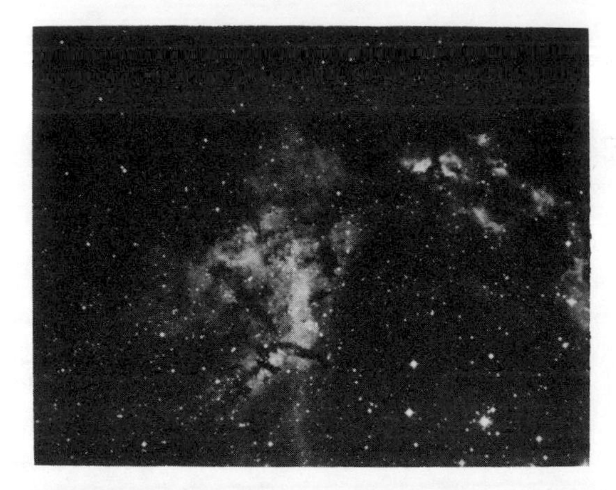

Fig. 15-4. The "Trifid" Nebula in Sagittarius: photographed in red light by the 200-inch telescope. The dark regions are caused by interstellar dust clouds between the nebula and the center, which absorbs visible light coming from the nebula. However, the microwave radiation emitted by molecules such as formaldehyde is not absorbed and so these molecules can be detected by radiosisotopes. (From Sagan, 1966.)

molecules in space. Formaldehyde, the prebiotic precursor of glycine and the sugars, is also present in large amounts. The discovery of cyanoacetylene in space is even more surprising. This molecule is relatively unstable and has not been studied very extensively by organic chemists. It was only a few years ago that it was proposed as the major precursor of the pyrimidines cytosine and uracil.

Little is known about the mechanism of formation of these substances. They are found in association with interstellar dust clouds, and it is generally believed that dust particles play some role in the synthesis of organic molecules. Experimental work designed to examine the formation of organic molecules from very dilute gases in the presence of solid particles is now beginning. Perhaps it will throw light on the way in which organic material is formed in space.

In the meantime, the relevance of these discoveries to the problem of the origins of life is unclear. The discovery that so many proposed prebiotic precursors are found in space lends some support to our ideas about prebiotic synthesis. However, we do not know whether organic chemicals formed in a dust cloud could later collect on the surface of planets accumulating within that dust cloud. It appears doubtful that the interstellar organic molecules could survive collision with a planet, but this is still an open question.

LIFE IN OUR SOLAR SYSTEM

The presence of life on Mercury is unlikely because the planet is very hot and has no atmosphere. Life on Uranus, Neptune, and Pluto is unlikely for the opposite reason—these planets are probably much too cold (see Table 2-1 for the temperatures of these planets).

The temperatures in the upper atmospheres of Jupiter and Saturn are also too cold to support life. However, the temperature must increase as one descends through the atmosphere, provided the atmosphere is not isothermal. If the atmosphere is adiabatic to a sufficient depth, there will be a region where the temperature is suitable for life. Even this does not mean that life is likely to be present.

Jupiter and Saturn are made up mostly of hydrogen and helium so there can be no solid surface on these planets. It is difficult to visualize how life could arise without a surface for accumulation and concentration of organic materials, but perhaps it is possible. If life did somehow get started, the organisms which evolved would need some mechanism to ensure that they stayed at a constant level in the atmosphere. If they rose too high, they would freeze and their growth would stop but they would not be killed. If they fell too far, the high temperatures would kill them. For these reasons we believe that life is very unlikely on Jupiter or Saturn. However, it is possible, but not likely, that organisms can control their density so that they remain at the same level in the atmosphere.

Uranus and Neptune contain much less hydrogen and helium than Jupiter and Saturn and presumably have a solid surface not too deep in the planet. However, they are even colder than Jupiter and Saturn, so life seems rather improbable on them.

The presence of methane and ammonia on Jupiter, and methane on Saturn, has been observed spectroscopically; their abundances are probably not more than a few percent of the H_2 and He abundances. Water is presumably there also, but it is not seen spectroscopically because it is frozen out lower in the atmosphere. Ultraviolet light from the sun and possible electric discharges are probably producing numerous organic compounds on these two planets (as well as on Uranus, Neptune, and Pluto). The identification of such compounds would be a very valuable project for a space probe.

Venus has long been thought of as a twin planet to the earth. It is about the same size and is covered with clouds. In the last 20 years, however, investigations have shown that Venus is different from Earth in many ways. Although there is a small amount of water in the atmosphere, the observed clouds are apparently not water, but consist of some as yet unidentified substance. The atmosphere is largely carbon dioxide, and the surface pressure is very high: 90 atmospheres, according to the latest Russian measurements. A temperature at the surface of about 500°C was first estimated by measuring the microwave emission spectrum. This value has been confirmed by measurements from a Russian instrument that landed on the surface. This most recent value of 477°C is consistent with an adiabatic atmosphere and the high surface pressures.

Life as we know it cannot exist at temperatures as high as 500°C. However, as in the case of Jupiter and Saturn, if the surface is very hot and the tops of the clouds are very cold (−40°C), there must then be a level in the atmosphere that is at a favorable temperature for life. All the problems of evolution and maintaining life which we emphasized in our discussion of Jupiter and Saturn apply equally to Venus. So although the possibility cannot be completely discounted, we think life very unlikely to be present.

LIFE ON MARS

The one planet in our solar system other than the earth on which life seems at all likely to occur is Mars. The history of speculations concerning life on Mars is unfortunate. In the 19th century, a group of astronomers claimed that they saw "canals" on the planet. Since the "canals" shifted from year to year, it was concluded that intelligent beings must be digging these "canals." A great deal of wild speculation was devoted to the nature of these intelligent beings. These publications fell so far short of normal scientific standards that planetary astronomy earned a bad reputation among most astronomers.

The 19th century observations on Mars were all visual rather than photographic. The human eye with the aid of a medium size telescope can resolve more detail than is seen in a photograph taken with one of the largest telescopes. However, the eye has a tendency to convert a series of dots or markings into a solid line, and the canals were mostly visual artifacts due to this unfortunate tendency to see straight lines where none exist. The most decisive evidence against the presence of "canals" or anything like them comes from the photographs taken from the Mariner 4, 6, and 7 spacecrafts that passed within a few thousand kilometers of Mars, as well as the orbiting spacecraft Mariner 9. Nothing like a "canal" can be seen in any of these photographs, although some more or less linear features are present.

The physical and chemical data obtained from the Mars flybys (as well as observations from the earth) show that the atmospheric pressure is between 5 and 10 millibars (1 atm = 1,013 millibars). The atmosphere is mostly CO_2. It is claimed that less than 1% of nitrogen is present, but this figure is still somewhat controversial. Water is definitely present in the atmosphere, but the amount is very small, corresponding to the vapor pressure of ice at −80°C (5×10^{-7} atm). The Martian icecap is apparently mainly solid CO_2 (dry ice) rather than water ice. But there is a good deal of water, and also some of the interesting carbon dioxide clathrate hydrate ($CO_2 \cdot 6H_2O$), in the icecap.

The surface temperatures on Mars are not extreme. At the equator, they vary from 175°K to about 305°K, and at the poles they range from 145°K to 265°K. The sublimation of the icecaps is associated with one of the most intriguing phenomena known to occur on Mars—the spreading of *waves of darkening*. Each year, dark zones appear close to the polar icecaps and then move gradually toward the equator. It has often been supposed

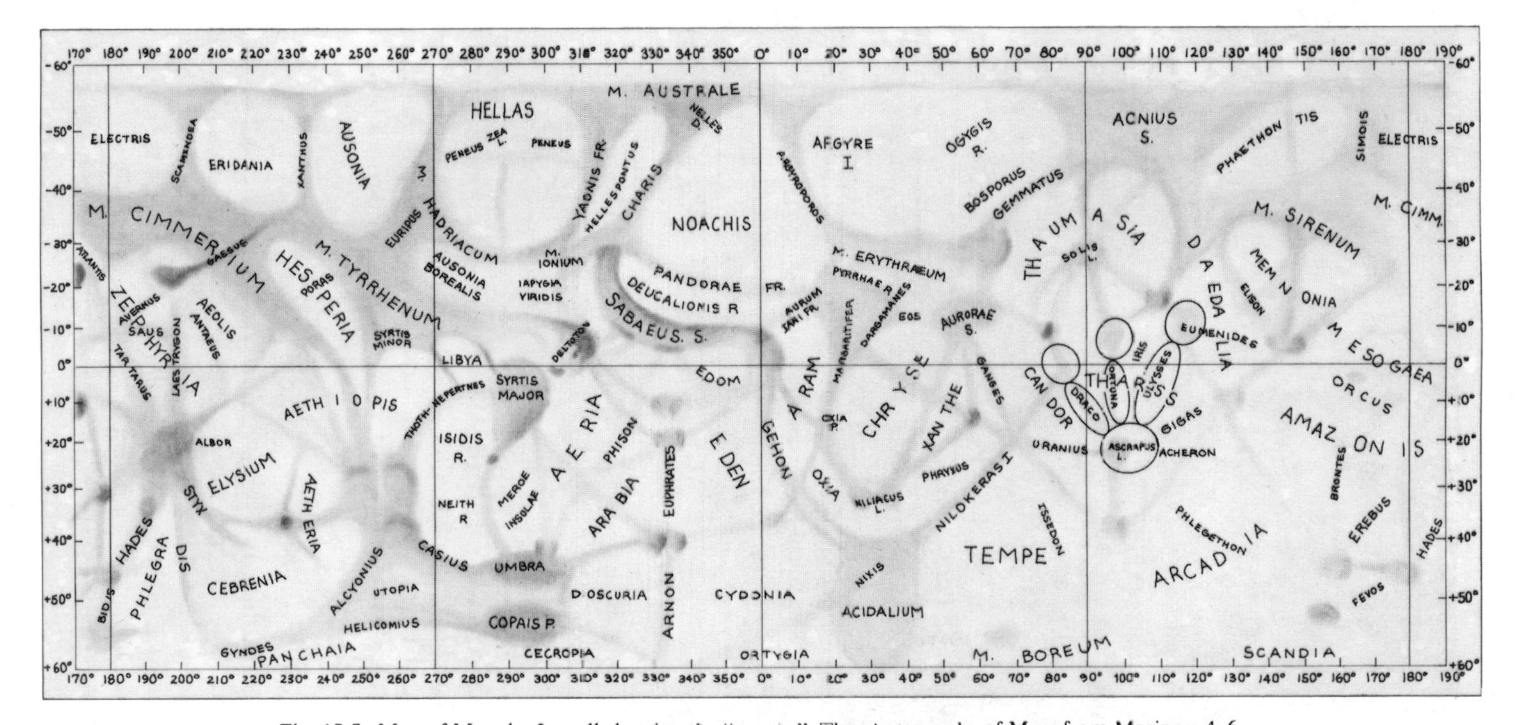

Fig. 15-5. Map of Mars by Lowell showing the "canals." The photographs of Mars from Mariners 4, 6, and 7 show no evidence at all of canals. The "canals" seen by Lowell are thought to be a visual artifact of the human eye.

Fig. 15-6. Mariner 6, photo 35, showing the planet Mars with the south polar ice cap at bottom. The ice cap consists mainly of solid CO_2 (dry ice). This picture was taken from a distance of 537,000 km on July 30, 1969. (Courtesy of National Aeronautics & Space Administration.)

that these changes in surface reflectivity are caused by the growth of vegetation, but there are many alternative interpretations. The changes could, for example, be due to the precipitation of water released from the polar icecaps or to the redistribution of dust by wind storms. Spectroscopic studies of the light reflected from Mars suggest that the red color of the planet is due to the presence of iron-containing minerals on its surface.

Photographs of the surface of Mars have been taken by three flybys—Mariner 4 in July, 1965, and Mariners 6 and 7 in July and August, 1969. The Mariner 9 orbited Mars on November 14, 1971 and took pictures for over a year. Estimates of the likelihood that there is life on Mars have changed drastically after each mission. Mariner 4 photographs showed a highly cratered surface. The number of craters was sufficiently high to convince most workers that there never had been an ocean or a large amount of water on Mars. If there had been, the craters would have weathered and eroded away.

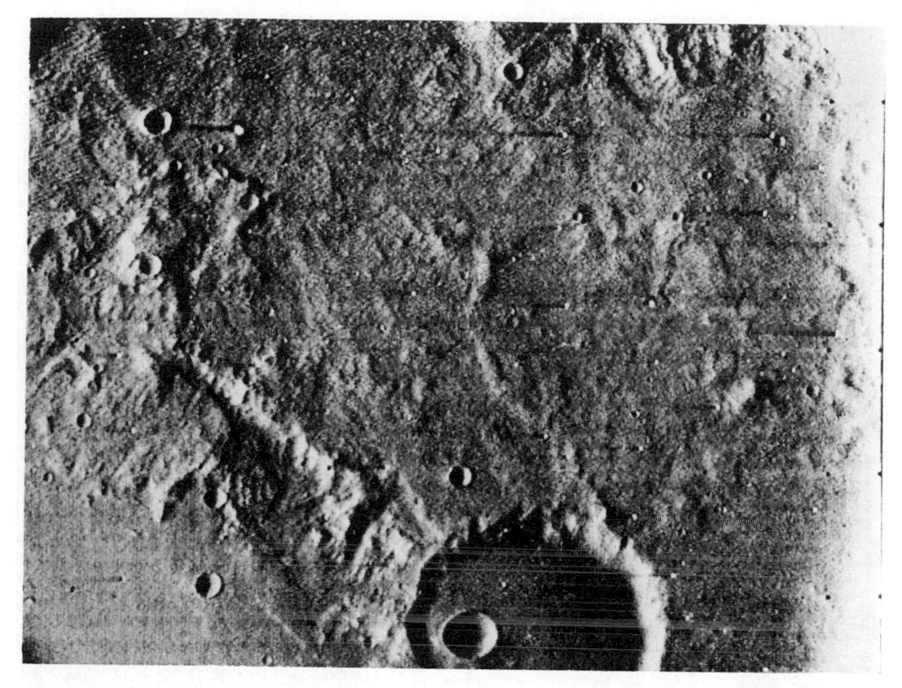

Fig. 15-7. Mariner 6, photo 21, showing the craters on the surface of Mars. The area shown here is 690 by 900 km (430 by 560 miles). The largest crater shown is about 260 km in diameter. This picture was taken from a distance of 3,460 km on July 30, 1969. (Courtesy of National Aeronautics & Space Administration.)

The photographs from Mariners 6 and 7 showed areas with large numbers of craters but also areas with a far smaller density of craters. These flybys showed that the icecaps were solid CO_2. Water was not detected in them although it was thought likely that some water was present. At that time, it was generally held that there never had been much water on Mars.

The Mariner 9 results came as a great surprise. At first, an enormous dust storm obscured the surface of the planet, but when the dust storm cleared, cratered areas were again seen. Some of the craters are impact craters (from large meteorites), but some of the craters are volcanic in origin. Since volcanoes, at least on the earth, emit large amounts of water as well as lava, these observations imply that there is considerably more water present than had previously been thought. Photographs of the south polar region showed smooth areas which are interpreted as being the result of the freezing and thawing of ground water. One of the most interesting photographs showed a meandering canyon (or channel) with branches. This is believed to have been formed by flowing water, although it could just possibly have been created by wind erosion or some other process.

Even with the evidence for a substantial abundance of water on Mars, these observations are not particularly favorable to the view that life exists

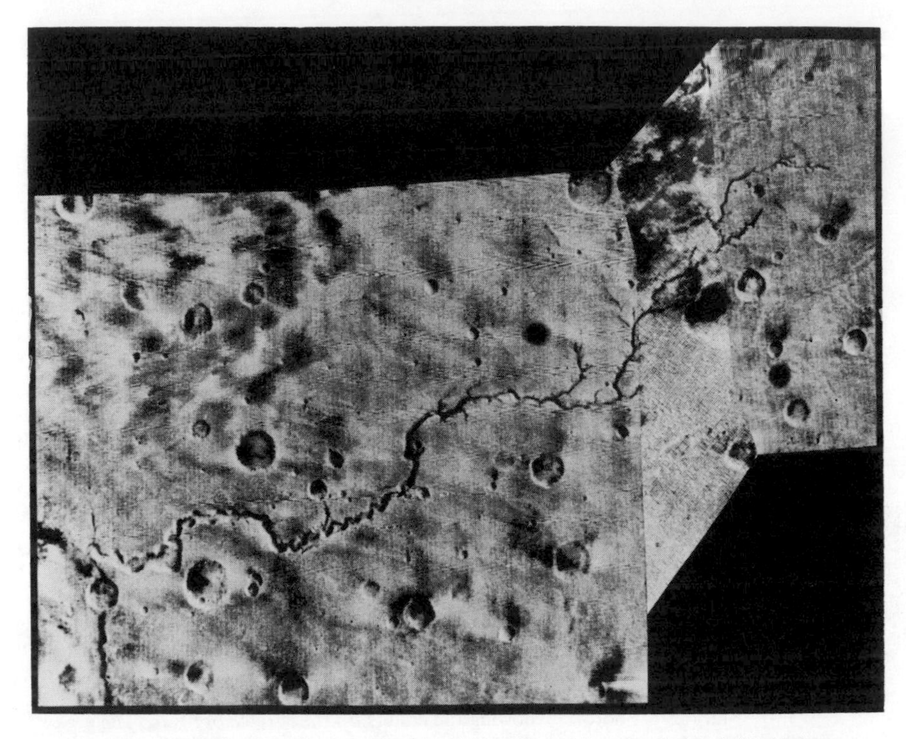

Fig. 15-8. Though not unique on the Martian surface, this meandering "river" photographed by Mariner 9 is the most convincing evidence that a fluid once flowed along the surface of Mars, draining an extended area and eroding a deep channel in the familiar manner. The feature, some 575 km (355 miles) long and 5 to 6 km (3 to $3\frac{1}{2}$ miles) wide, resembles a giant version of an earth "arroyo"—a watercut gully found frequently in the mountainous southwestern United States. Scientists believe, however, that not nearly enough water exists in the Martian atmosphere to allow formation of rivers. Mariner 9 infrared spectral data, as well as earth-based instruments, show very little water on Mars at the present time. The Martian valleys also resemble sinuous rilles on earth's moon believed to be associated with lava flows. However, no lunar rilles display the branching tributaries seen in the Martian valleys. This feature, first seen on January 19, 1972, extends from about 38° to 45° west longitude and 27° to 30° south latitude in the mosaic. (Courtesy of Jet Propulsion Laboratory, California Institute of Technology; and National Aeronautics & Space Administration.)

there. The temperatures are not unfavorable; many earth organisms can survive the repeated freezing and thawing that would occur on Mars. The main problem is still the shortage of available water either in the atmosphere or loosely absorbed in minerals (this can also be viewed as a shortage of hydrogen). If organisms are to grow on the surface of Mars at 25°C, then they must accumulate water when the relative humidity is $10^{-4}\%$. This would be difficult but perhaps not impossible. If the organisms were to grow at -80°C, the relative humidity would be 100% and there would then be little difficulty in accumulating water. Organisms growing under these

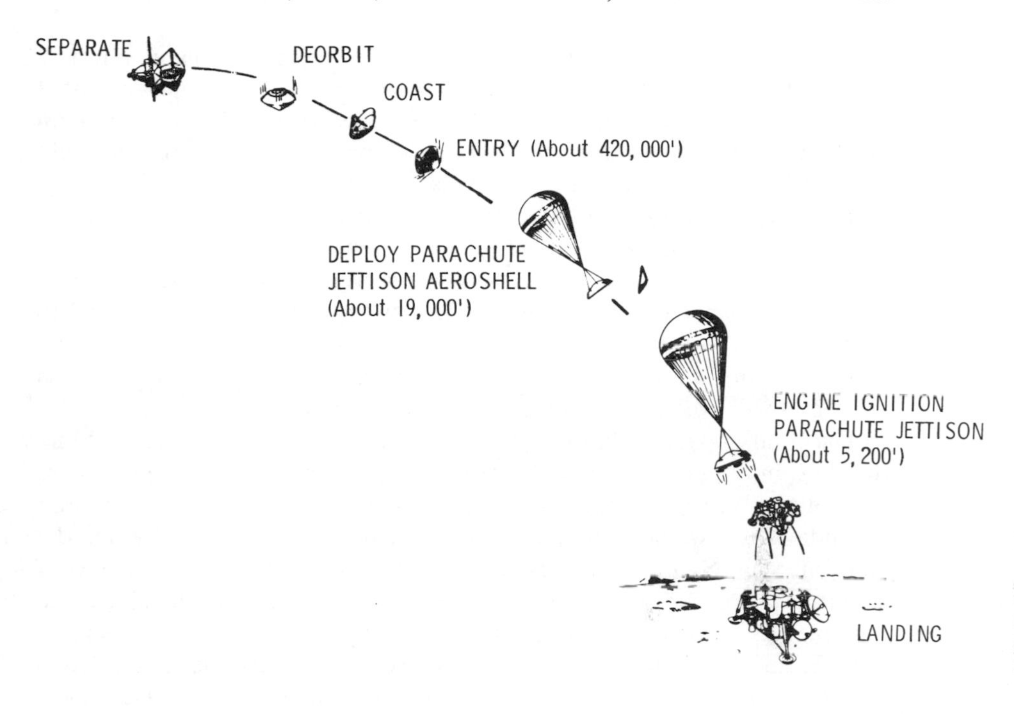

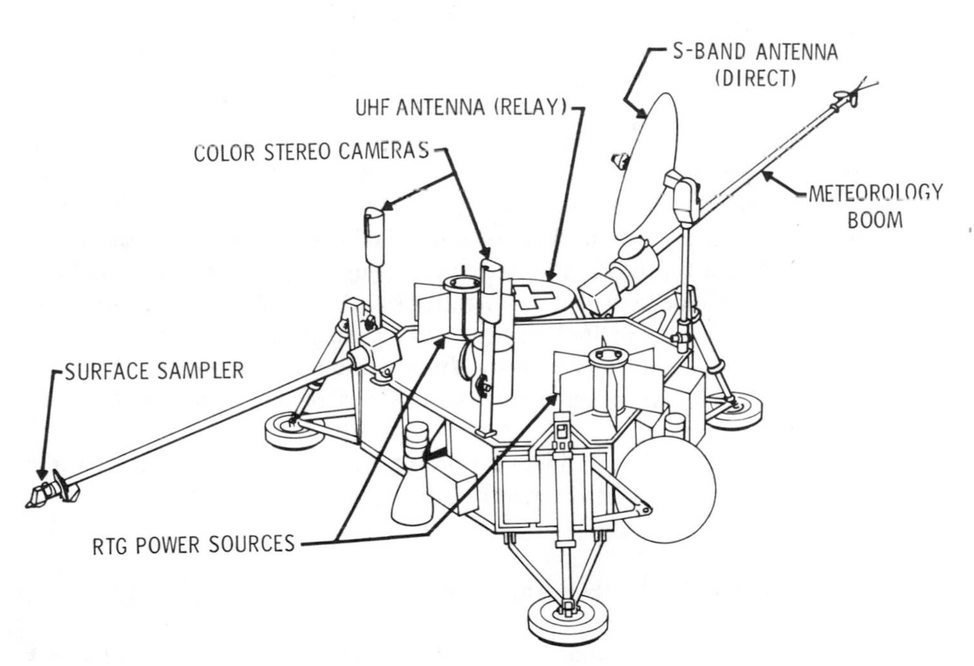

Fig. 15-9. The 1975 Viking mission to Mars. Shown here are the lander entry sequence and the landed configuration of the spacecraft. The Viking mission will conduct experiments on the surface of Mars to determine whether life is present on that planet.

conditions would have to contain a very concentrated salt or organic solution, but this may be possible since a saturated solution of LiCl, a supersaturated $CaCl_2$ solution, or a 67 wt% C_2H_5OH solution (Fig. 9-1) would remain liquid at −80°C. Of course, we do not know whether organisms could adapt and grow at temperatures as low as −80°C.

The above discussion is concerned with the survival of life on Mars. We want to emphasize that life may have arisen on Mars in the past when large amounts of liquid water may have been present for times long enough for a living organism to arise. The organisms to survive would have had to adapt to the relatively unfavorable conditions as the planet dried out. We see no fundamental difficulty in this. However, if Mars has always been as dry as it now is, we doubt that life could have ever arisen.

The total apparent absence of nitrogen from the atmosphere of Mars would be an argument against the presence of life, because if organisms are metabolizing nitrogen compounds such as amino acids, or purines and pyrimidines, we would expect that some of the nitrogen would enter the atmosphere as N_2 molecules. However, organisms might easily survive if the atmosphere contained 0.1% nitrogen, a value not excluded by the available data. We know that on the earth, 0.04% CO_2 in the atmosphere supports an abundant biosphere. Alternatively, it is possible that there is nitrogen in the form of nitrates in the soil. Then life might be present even though the amount of nitrogen in the atmosphere is small. However, it is possible that there never has been a significant amount of nitrogen on Mars. In that case, life based on proteins and nucleic acids could not have arisen.

The numerous uncertainties discussed above make it impossible to estimate the probability that life is present on Mars. Our knowledge of this planet is still very slight. We suspect that the Viking Mission, which will land two packages of about 50 lb of scientific instruments on the surface of Mars in 1976, will produce many surprises, just as previous missions have done. The instruments will include a mass spectrometer attached to a gas chromatograph that will analyze the organic constituents of the soil. There will also be a series of experiments designed to detect the presence of living organisms. These experiments involve a series of assays for growth and metabolic activity.

The Viking Mission is, in our opinion, the most important scientific experiment that NASA has carried out or is planning to carry out.

LIFE OUTSIDE THE SOLAR SYSTEM

Life at all similar to that on the earth could only survive at moderate temperatures and presumably only on a planet. An organism whose biochemistry is based on carbon and water could not survive at temperatures much above 100°C, because familiar biopolymers would be hydrolyzed rapidly. Even an unfamiliar biochemistry based on carbon could hardly operate above 500°C since most carbon compounds pyrolyze at that temperature.

It is unlikely that organisms could grow at temperatures below −100°C because the rates of important chemical reactions would become too slow, and because there are few solvents that remain liquid below this temperature (ethyl alcohol-water is liquid to −126°C—see Fig. 9-2). Some terrestrial organisms can be stored at liquid nitrogen temperatures (−195°C) or even liquid helium (−269°C), but no organisms grow at temperatures much below −15°C.

These considerations would not apply to forms of life based on liquid ammonia or other solvent systems which might function at lower temperatures. Life forms based on silicates might operate at higher temperatures. Speculations about such systems are frequently brought up at cocktail parties and that is where they belong.

In deciding whether life is abundant elsewhere, it is necessary to estimate whether suitable environments for life, that is planets with moderate climates, are abundant. Unfortunately, a planet the size of the earth associated even with one of the nearest stars would not be observable using present-day astronomical methods. We have therefore to estimate the number of suitable planets indirectly.

Most modern theories of the formation and evolution of stars predict that planets should be relatively common. In our solar system there are nine planets and hundreds of asteroids. Although a planet the size of the earth or even Jupiter could not be observed on a nearby star, a number of larger, nonradiating bodies have been observed. Table 15-3 lists these objects. The smallest object is the companion to Barnard's Star which is only 50% greater in mass than Jupiter. Single, double, and triple star systems are quite abundant. Table 15-4 gives their frequency.

Table 15-3. Invisible Stellar Companions[a,b]

Star	Fraction of the Sun's Mass
Barnard's Star	~0.0015
Lalande 21185	~0.01
61 Cygni	≥0.008
Krueger 60A	~0.009–0.025
Bd + 20° 2465	≥0.02
Ci 2354	≥0.02
η Cas	~0.01

[a] From H. Brown, *Science* **145**, 1177 (1964).
[b] The mass of Jupiter is 0.001 of the mass of the sun.

Table 15-4. Frequency of Stellar Systems within 20 Parsecs[a,b]

	Observed Number of Systems	Observed Individual Stars
Single	746	746
Double	159	318
Triple	10	30
Quadruple	0	0

[a] From H. Brown, *Science* **145**, 1177 (1964).
[b] One parsec is equivalent to 3.26 light years.

It is possible to work out the statistical distribution of the masses of objects formed in the condensation of dust clouds. When the parameters in these theories are adjusted to fit the known distribution of single, double, and triple stars, the theories predict that most star systems include planets.

For a planet to be able to support the evolution of life, the temperature of the planet should remain in the range $-80°C$ to $+100°C$ for at least 100 million years. Thus, the distance of the planet from its sun, the energy radiated by its sun, and the time for which the rate of energy production is roughly constant are all important. The energy production of a star is related to its size, the large stars being more luminous. Figure 15-10 shows a Hertzsprung-Russell diagram in which the luminosity of a star is plotted against its spectral type. The surface temperatures derived from the spectral type observations are included in Fig. 15-10, along with the time for which each type of star maintains approximately constant luminosity.

The path of our sun on this diagram can be estimated theoretically. It stays on the main sequence for about 10^{10} years, after which all the hydrogen is "burned" up to helium and it becomes a white dwarf. On the other hand a star of spectral class BO would have a mass 17 times that of our sun and would radiate 30,000 times as much. Its path on the Hertzsprung-Russell diagram would be very different from that of our sun, and such a star would remain on the main sequence for only eight million years. Since this is presumably insufficient time for the origin of life and its extensive evolution, the planets surrounding such stars are considered unsuitable for advanced forms of life.

Stars larger than AO have lifetimes on the main sequence of less than 4×10^8 years, so it is not likely that planets surrounding such stars have highly developed life on them. The M stars, on the other hand, should remain on the main sequence for as much as 100×10^9 years. However, their luminosity is very low, and the habitable zone for planets is quite narrow. So it is felt that M stars are not particularly likely to have planets with life on them. This leaves us with stars between AO and K5 that are particularly suitable for life. This is about 10% of all stars.

Another factor has to be considered. The orbits of planets associated with double stars may be such that temperature on the planet changes very

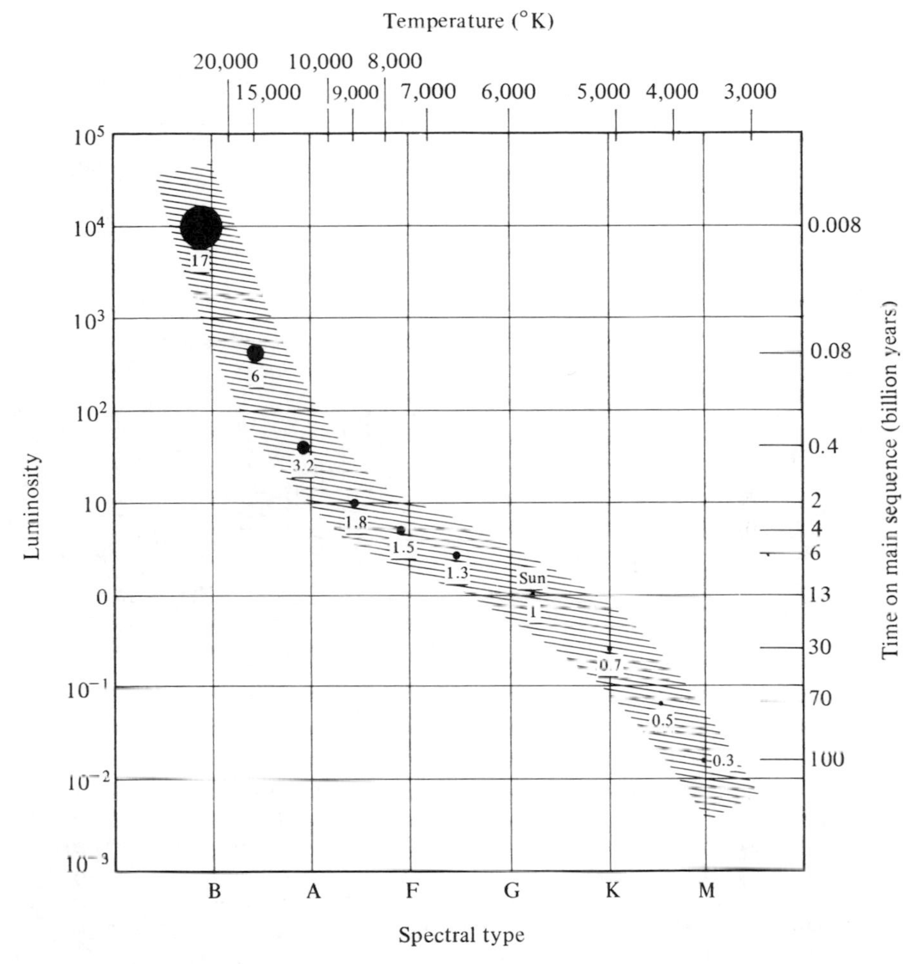

Fig. 15-10. The Hertzsprung-Russell Diagram (schematic drawing). When stars are classified by brightness and spectral type (temperature), most lie in a band known as the "main sequence." Their temperature is determined primarily by their mass. The heavier the star, the hotter, brighter, and shorter-lived it is. Luminosity, on the left, is shown on a logarithmic scale with the brightness of our sun as one. Temperature, across the top, is in degrees Kelvin. Star sizes are shown schematically, the figures below indicating mass, with the mass of the sun as one. A star whose spectral type is midway between F and G is called an F5. The sun is a G2 star.

greatly during a single revolution. This would probably make life difficult or impossible. A second difficulty is associated with the instability of planetary orbits around double stars. Many of these orbits would change greatly over periods of much less than 100 million years. Thus, only a few percent of binary star systems have planets moving in orbits suitable for the maintenance of life. Since binary stars are approximately half of all stars, this consideration excludes half of all stars as suitable centers for life.

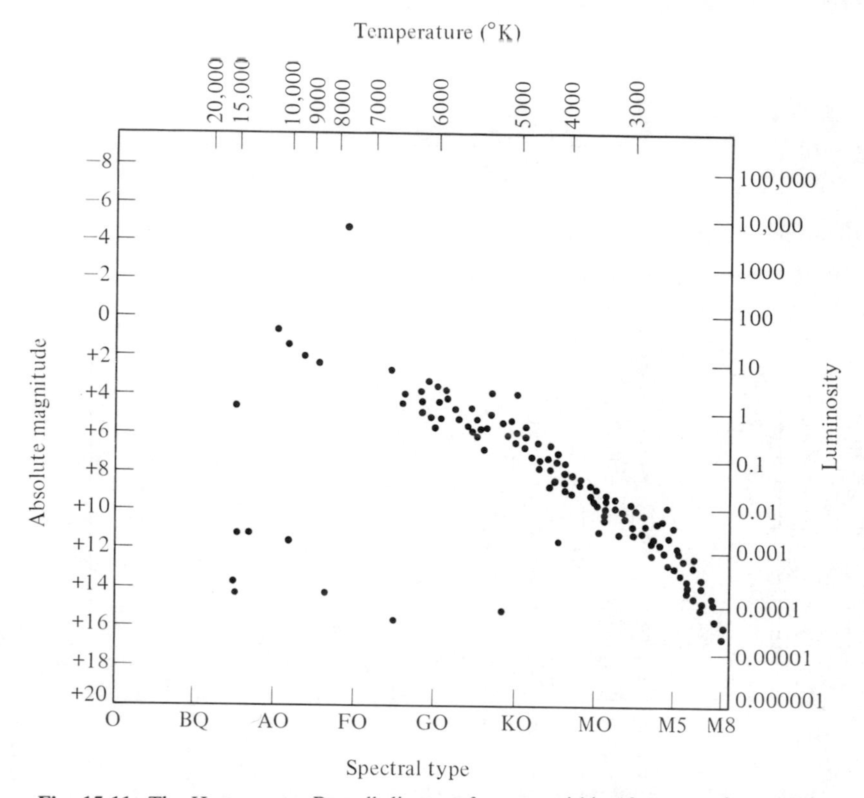

Fig. 15-11. The Hertzsprung-Russell diagram for stars within 10 parsec of sun. This distribution of star types is presumably representative of our galaxy and possibly other galaxies as well. The abundance of M and K stars shows up clearly. Stars below the main sequence are white dwarfs. Stars lying above the main sequence are red giants (more are located within 10 parsecs of the sun). A decrease of 5 in the absolute magnitude corresponds to an increase of a factor of 100 in the luminosity. (From Struve *et al.*, *Elementary Astronomy*.)

Life would not necessarily evolve on any planet which happened to have a suitable temperature. If Jupiter was at the same temperature as the earth, the hydrogen and helium would not be able to escape, because of the higher gravitational field. A planet of Jupiter's composition would therefore probably not be suitable for the evolution of life. Similarly, if a planet of the earth's temperature retained very little water or nitrogen when it condensed, it is unlikely that life would have evolved. This may have been the case with Venus and Mars. It is difficult to estimate the probability that life would be excluded for one or the other of these reasons.

Finally we should estimate the probability that, although the temperature and atmosphere were suitable and all the necessary organic compounds were synthesized, the organic polymers did not organize into a self-replicating system. We do not know enough about the origin of life to make any

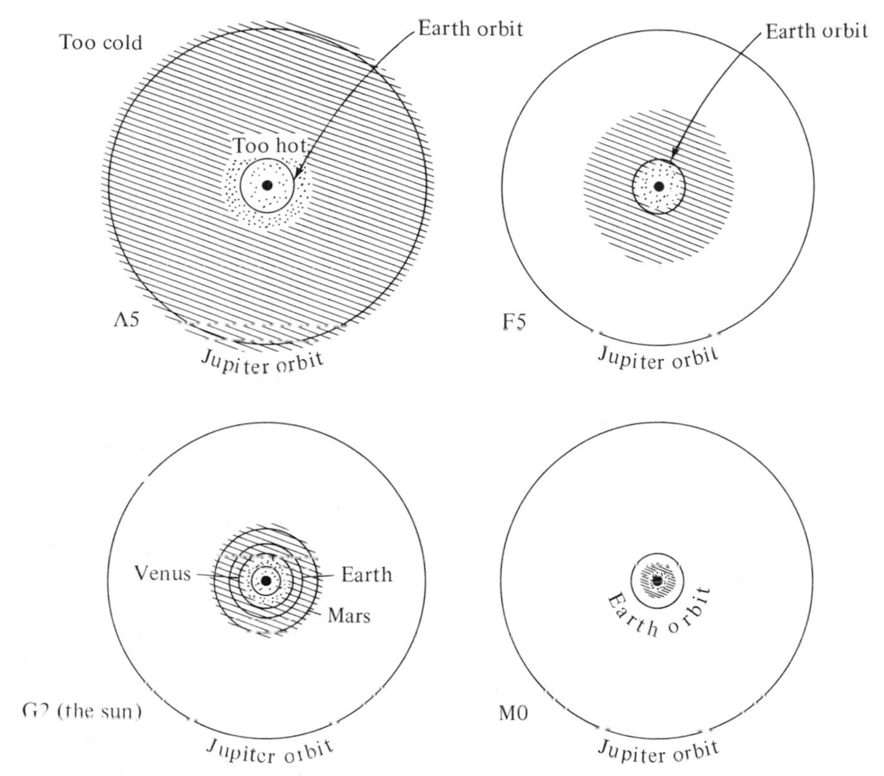

Fig. 15-12. Thermally habitable zone of various types of stars is represented by cross-hatched area. The A5 spectral type has largest zone, but the star does not remain stable long enough for the origin of life to take place on a planet near it. The habitable zone of our sun (a G2 type) extends from orbit of Venus to orbit of Mars. The habitable zone of a M0 star is so narrow that the probability of a planet being in this zone is small.

estimate of this probability. It is equally hard to discuss the probability that life, having originated, would have survived for a long time. In our view these are the weak points in the estimates of the abundance of life in the universe.

It appears very unlikely that life in other solar systems can be detected except by traveling to the other solar systems or by intercepting radio communication. Laser communication has also been considered seriously. Since space travel to other solar systems is unlikely in the foreseeable future, most attention has been placed on radio detection. Attempts have therefore been made to estimate the number of planets harboring civilizations advanced enough to communicate by radio.

The number of communicating civilizations (N) can be calculated by the following formulas:

$$N = N_* f_p n_e f_l f_i f_c$$

$$N = R_* f_p n_e f_l f_i f_c L$$

where

$N_* =$ Number of stars in observable universe
$R_* =$ Rate of star formation
$f_p =$ Average number of planets per star
$n_e =$ Fraction of planets with environments suitable for life
$f_l =$ Fraction of these planets on which life does arise
$f_i =$ Fraction of life-bearing planets on which intelligent life evolves
$f_c =$ Fraction of these planets which develop the technology needed to use radio communication
$L =$ Mean life of communicating civilizations

The first formula assumes that a communicating civilization, once formed, survives as long as the planet. The second takes account of the destruction of civilizations.

Table 15-5 gives two different estimates of these parameters. The first is a conservative estimate by Shapley who did not consider the rate of star formation or the rate of destruction of civilizations. The second estimate made by Sagan and Drake is much more optimistic. Also included in the table is space for the reader to put in his own numbers. These can be considered as reliable as the other two estimates.

There are about 10^{11} stars in our galaxy and 10^{20} in the universe. Using the "conservative" estimate there would be 10^8 planets with life and 10^2 communicating civilizations. Using the optimistic estimate, there would be 10^{10} or 10^{16} communicating civilizations in the universe, depending on the value of L that is chosen. Published estimates of the probable number of communicating civilizations *in our galaxy*, therefore, range from 10^{-7} to 10^9.

Table 15-5. Number of Communicating Civilizations

	Conservative Estimate	Optimistic Estimate	Reader's Estimate
R_*	—	$10^{10}\ yr^{-1}$ ($10\ yr^{-1}$ for a galaxy)	________
f_p	10^{-3}	1	________
n_e	10^{-6}	1	________
f_l	10^{-3}	1	________
f_i	$(10^{-3})^a$	0.1 to 1	________
f_c	$(10^{-3})^a$	0.1 to 1	________
L	—	10^2 to 10^8	________

[a] Shapley did not give numbers for f_i or f_c. We use 10^{-3} for these factors as a continuation of his conservative argument. Shapley was probably more optimistic than his estimate would indicate; he used these low values only to present the least favorable case.

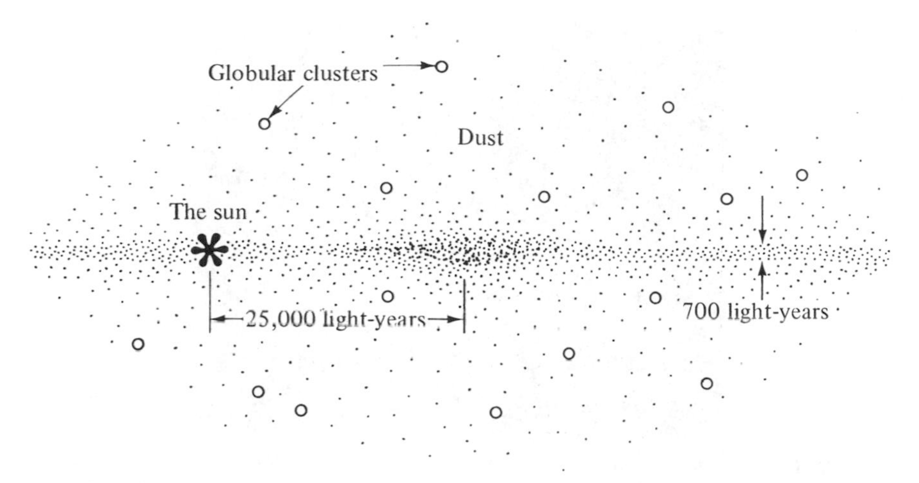

Fig. 15-13. Our galaxy seen edge-on. The milky way is seen when an observer looks toward the center of the galaxy. There are about 10^{11} stars in our galaxy. Outside the galaxy and in most parts of the galaxy, the vacuum contains only 1 hydrogen atom per cm^3.

Fig. 15-14. A typical galaxy seen face-on (NGC 5364). (From Mount Wilson and Palomar Observatories.)

Fig. 15-15. Eavesdropping on space with an 85-foot-wide dish antenna, a radio telescope in West Virginia, shown in a star-streaked time exposure, listens for radio signals from planets or stars during 1960's Project Ozma. Although the results of this project were negative, the experiment pointed to the day when improved equipment will permit communication by radio with other worlds. A new device, called a laser and employing a synthetic ruby to produce a pencil-thin beam of light, opens up the possibility of sending messages with pulses of light. (From Life Nature Laboratory, *The Earth.*)

In spite of the uncertainties in this numbers game, many responsible scientists take seriously the possibility that intelligent societies exist elsewhere and can be communicated with. A good deal of attention has been given to the problem of analyzing radio-transmissions to distinguish communication signals from random radio noise from space, and from coherent signals due to processes unconnected with intelligent civilizations.

So far, only a few hundred hours of radio-telescope time have been devoted to listening for such signals (Project Ozma). The results have been negative, as might have been expected. A program that would be likely to detect such signals would require extensive use of large and costly radio-telescopes. As yet, the funding for such a program has not been made available, but it seems likely to us that a large-scale search will be initiated within the next fifty years.

References

Meteorites

G. S. Hawkins, *The Physics and Astronomy of Meteors, Comets and Meteorites* (McGraw-Hill, New York, 1964).

H. H. Nininger, *Out of the Sky; an Introduction to Meteorites* (Dover, New York, 1959).

B. Mason, *Meteorites* (Wiley, New York, 1962).

J. A. Wood, *Meteorites and the Origin of Planets* (McGraw-Hill, New York, 1968).

Carbonaceous chondrites

J. M. Hayes, "Organic constituents of meteorites—a review," *Geochim. Cosmochim. Acta* **31**, 1395–1440 (1967). This review critically analyzes all the data available up to 1967 on organic material in carbonaceous chondrites.

B. Mason, "The carbonaceous chrondrites," *Space Sci. Rev.* **1**, 621–646 (1963).

E. K. Gibson, C. B. Moore, and C. F. Lewis, "Total nitrogen and carbon abundances in carbonaceous chondrites," *Geochim. Cosmochim. Acta* **35**, 599–604 (1971).

K. Kvenvolden, J. Lawless, K. Pering, E. Peterson, J. Flores, C. Ponnamperuma, I. R. Kaplan, and C. Moore, "Evidence for extraterrestrial amino-acids and hydrocarbons in the Murchison Meteorite," *Nature* **228**, 923 (1970).

K. A. Kvenvolden, J. G. Lawless, and C. Ponnamperuma, "Nonprotein amino acids in the Murchison Meteorite," *Proc. Nat. Acad. Sci. U.S.* **68**, 486 (1971).

C. E. Folsome, J. Lawless, M. Romiez, and C. Ponnamperuma, "Heterocyclic compounds indigenous to the Murchison Meteorite," *Nature* **232**, 108 (1971).

J. R. Cronin and C. B. Moore, "Amino acid analyses of the Murchison, Murray, and Allende carbonaceous chondrites," *Science* **172**, 1327 (1971).

J. G. Lawless, K. A. Kvenvolden, E. Peterson, C. Ponnamperuma, and C. Moore, "Amino acids indigenous to the Murray Meteorite," *Science* **173**, 626 (1971).

J. Oró, J. Gilbert, H. Lichtenstein, S. Wikstrom, and D. A. Flory, "Amino acids, aliphatic and aromatic hydrocarbons in the Murchison Meteorite," *Nature* **230**, 105 (1971).

K. L. Pering and C. Ponnamperuma, "Aromatic hydrocarbons in the Murchison Meteorite," *Science* **173**, 237 (1971).

J. G. Lawless, K. A. Kvenvolden, E. Peterson, C. Ponnamperuma, and E. Jarosewich, "Evidence for amino-acids of extraterrestrial origin in the Orgveil meteorite," *Nature* **236**, 66 (1972).

Comets

F. G. Watson, *Between the Planets* (Harvard Univ. Press, Cambridge, 1956).

N. B. Richter, *Statistik und Physik der Kometen* (J. A. Barth, Leipzig, 1954).

The moon

G. Eglinton, J. R. Maxwell, and C. T. Pillinger, "The Carbon Chemistry of the Moon," *Scientific American* **227**, 80 (Oct. 1972).

Interstellar molecules

L. E. Snyder and D. Buhl, "Molecules in the interstellar medium," *Sky* and *Telescope* **40 (5)**, 1 (1970); **40 (6)**, 5 (1970); **45 (3)**, 156 (1973); *Nature*, **234**, 332 (1971).

D. M. Rank, C. H. Townes, and W. J. Welch, "Interstellar molecules and dense clouds," *Science* **174**, 1083 (1971).

Life in our solar system

P. Lowell, *Mars and its Canals* (Macmillan, New York, 1906); *Mars as the Abode of Life* (Macmillan, New York, 1908). These volumes show numerous drawings of Mars together with speculation about an advanced civilization on that planet.

C. Ponnamperuma and H. P. Klein, "The coming search for life on Mars," *Quart. Rev. Biol.* **45**, 235–258 (1970).

H. Morowitz and C. Sagan, "Life in the clouds of Venus?" *Nature* **215**, 1259 (1967).

C. Sagan, "The solar system beyond Mars: An exobiological survey," *Space Sci. Rev.* **11**, 827–866 (1971).

Life outside our solar system

I. S. Shklovskii and C. Sagan, *Intelligent Life in the Universe* (Holden-Day, San Francisco, 1966). This is the best discussion in this area with particular emphasis on the astronomical aspects of the problem.

A. G. W. Cameron, ed., *Interstellar Communication* (W. A. Benjamin, New York, 1963).

This volume contains many of the original papers relevant to this topic.

There are many popular books in this area. Included are:

H. Shapley, *Of Stars and Men* (Beacon Press, Boston, 1958).

W. Sullivan, *We are not Alone* (McGraw-Hill, New York, 1964).

F. Hoyle, *Of Men and Galaxies* (Univ. of Washington Press, Seattle, 1964).

Chapter Sixteen

Summary and Outstanding Problems

It is almost certain that all terrestrial forms of life are descended from simple, single-cell organisms that evolved 3 billion or more years ago on the earth. We have described in some detail the stages that we believe led to the formation of these earliest organisms from the inorganic constituents which made up the surface of the primitive earth.

The solar system was formed about 4.5 to 5.0×10^9 years ago from a diffuse dust cloud. The central part of the dust cloud condensed into the sun, while the outer parts formed the planets. The temperature in the dust cloud at the time when the planets were condensing fell off rapidly with increasing distance from the sun. It is unlikely to have exceeded 300°C in the region where the earth was formed.

The cosmic dust cloud must have contained a great deal of hydrogen, but this gas was lost by the earth during the condensation process. Since much of the neon and argon from the dust cloud was lost at this time as well, we must suppose that free methane, ammonia, and nitrogen would also have escaped from the earth's gravitational field. It follows that most of the carbon, nitrogen, and oxygen on the surface of the earth must have been retained as nonvolatile compounds when the earth was formed.

Organic compounds may well have been abundant in the original dust cloud, but they would have been distributed uniformly throughout the earth. While it is uncertain whether the whole earth melted immediately after its formation, the interior must have been maintained at a high temperature, as it is today, by the heat evolved from the decay of radioactive elements. Most of the carbon and nitrogen that is present at the surface of the earth must have been volatilized from the interior at that time.

The classical fossil record dates from the early Cambrian period and covers about 600 million years. Living organisms were present on the earth long before the Cambrian period, but they have not been investigated in detail because they did not leave macroscopic fossils. The systematic study of Precambrian microfossils has just begun. We already know that microorganisms that resemble modern bacteria and algae were present on the earth 3 billion years ago. It is almost certain that these organisms contained many of the biochemicals present in modern cells.

We know virtually nothing about the evolution of life in the period between the formation of the earth 4.5×10^9 years ago and the time when the earliest known fossils were formed. It is impossible at present to decide whether life evolved soon after the formation of the earth or whether a very long period of prebiological organic synthesis preceded the evolution of the simplest organisms. Clearly, it is most important to search for older sedimentary rocks to see whether or not they contain microfossils. Unequivocal evidence should be sought showing that some of the organic compounds found in association with early Precambrian organisms were laid down at the same time as the fossils.

The atmosphere of the earth soon after its formation must have been much more reducing than the present atmosphere. It is certain that no free oxygen could have been present. The atmosphere may have contained some ammonia, but nitrogen was most likely present predominantly as N_2. The main uncertainty concerns the form of carbon in the atmosphere. One extreme view is that all of the carbon in the primitive atmosphere was present as methane; the other extreme view proposes that the earth's atmosphere was formed by the degassing of CO_2. It is not at present possible, on the basis of the available geochemical evidence, to decide which of these views is closer to the truth.

We are prejudiced and believe that the primitive atmosphere was strongly reducing, but the major arguments present in Chapter 7 on organic synthesis would not be invalidated if it turned out that the atmosphere was less reducing than we think. The reactions that we describe would go on in an atmosphere containing CO_2, CO, and H_2, or in an atmosphere containing CO_2 and H_2, but no other form of carbon. In the light of recent experiments demonstrating the formation of formaldehyde from a simulated Martian atmosphere, it is clearly desirable to study the formation of a wide range of organics from a weakly-reducing atmosphere.

The simple reactive molecules that took part in further organic syntheses were formed in the atmosphere, mainly by the action of electric discharges and ultraviolet light. Enough energy was available to produce very large amounts of organic material in periods of 10^6 to 10^9 years. Hydrogen cyanide, formaldehyde, aldehydes, nitriles, and acetylenes were particularly important intermediates. While some larger molecules may have been formed in the atmosphere, the synthesis of most of the more complicated organic molecules probably occurred in aqueous solution, or in material deposited from aqueous solutions. Presumably, the oceans were the places where the bulk of the reactions occurred, but water droplets in the atmosphere and lakes may also have been important sites for some organic syntheses.

The oceans today are maintained at a pH of about 8.1 by a series of complex equilibria involving clay minerals. It is probable that the pH of the primitive ocean was controlled in a similar way and was therefore close to 8.0. The concentration of ammonia in the primitive ocean is likely to have been in the range of 0.01 to 0.001 M. Many of the prebiotic syntheses which we describe take place satisfactorily at pH 8.0 in the presence of 0.01 M ammonium ion.

A number of prebiotic syntheses of the naturally occurring amino acids from simple starting materials are known. It is a very striking fact that as much as 5% of the carbon in a mixture of methane, ammonia, and water can be converted into amino acids by the action of an electric discharge. At the present time, about 17 of the 20 amino acids that occur in proteins have been obtained under plausibly prebiotic conditions.

The synthesis of sugars from formaldehyde was described a century ago by Butlerov. This reaction has been reinvestigated repeatedly and has been carried out at moderate temperatures and pH's on the surface of alumina and carbonate-apatite. The product mixture contains some ribose and glucose, along with a large number of other sugars.

One of the most remarkable prebiotic reactions is the formation of adenine from hydrogen cyanide. This synthesis proceeds with up to 30% yield in liquid ammonia or with more modest yield under prebiotic conditions. Guanine can be made in a very similar way. Cytosine is formed from cyanoacetylene and cyanamide or cyanate. In aqueous solution cytosine is hydrolyzed to uracil and ammonia.

Most of the components of the genetic system have been synthesized from very simple starting materials under prebiotic conditions, but a number of problems remain. The different reactions that have been proposed are often incompatible. Thus, hydrogen cyanide can polymerize to adenine, and formaldehyde can condense to form sugars, but a mixture of the two in the presence of ammonia would give glycine. This may not be serious, since many fractionation processes might have occurred in the primitive earth. Many detailed kinetic studies will be needed before we can hope to formulate a consistent scheme for the synthesis of the thirty-odd molecules needed to build the first genetic system.

The instability of many important biochemicals presents a much more difficult problem. The hydrolysis of glutamine to glutamic acid, or of cytosine to uracil, would have been completed in quite a short time. Similarly, the alkaline hydrolysis of sugars is rapid on the geological time scale. It is not known how such compounds managed to accumulate on the primitive earth. Perhaps they were preserved as dry solids or were repeatedly reformed from their hydrolysis products.

The prebiotic accumulation of many important monomers may well have occurred in the bulk of the oceans. In some cases, however, it seems almost certain that higher concentrations must have been needed than could have built up there. It also seems likely that many of the condensation reactions in which polypeptides and polynucleotides were formed could only have occurred in concentrated solution or in the solid state. Thus, concentration mechanisms must have been important for the origins of life.

The most obvious method of concentrating dilute aqueous solutions of organic compounds is by evaporation. Lakes and tidepools are places where this could have occurred. Adsorption on minerals may also have been significant. A concentration mechanism that is possible even when the solute is volatile involves freezing: A dilute aqueous solution deposits ice when cooled below 0°C and the solute concentrates until a saturated solution is obtained. We think it likely that many important stages in the evolution of life, particularly the later ones, occurred at low temperatures.

The role of membranes during the earliest stages in the evolution of life is unclear. Once membranes had formed, they would certainly have been able to perform a useful function by holding together sets of prebiotic molecules, including polypeptides and polynucleotides. However, it is not clear that robust and stable membranes could have formed prebiotically, and it seems possible that the original concentration of prebiotic polymers occurred on the surface of solid particles (e.g., hydroxylapatite particles) or in colloidal droplets (coacervates). We expect that as our knowledge of the structure and properties of biological membranes increases, more efforts will be made to characterize potentially prebiotic membrane systems.

Although many important biochemicals can be synthesized under prebiotic conditions, we know much less about prebiotic condensation reactions. It has, for example, turned out to be surprisingly difficult to obtain nucleosides from bases and sugars.

A number of prebiotic syntheses of nucleotides from nucleosides and inorganic phosphate have been proposed. Cyanamide and a number of related compounds bring about the condensation of nucleosides and inorganic phosphate in aqueous solution to give a mixture of isomeric nucleotides. All of these reactions are inefficient, even in concentrated solutions of inorganic phosphate. Nucleosides can also be phosphorylated by heating with an acidic inorganic phosphate, such as $Ca(H_2PO_4)_2$, but these latter compounds are not readily formed under prebiotic conditions. A dry mixture of ammonium chloride, urea, and inorganic phosphate is a very effective phosphorylating agent which could have been present on the primitive earth.

The condensation of amino acids to peptides in aqueous solutions is another reaction that has been studied repeatedly but with rather inconclusive results. Cyanamide and similar reagents bring about the reaction in aqueous solution, but the yield is always poor under prebiotic conditions. Trimetaphosphate is a more effective condensing agent which could possibly have been present on the primitive earth. Thermal syntheses of polypeptides have also been studied extensively. It is unlikely that they could have taken place at temperatures much above 100° on a large scale.

A satisfactory synthesis of polynucleotides from monomers is lacking. Poor yields of dinucleotides can be obtained in solution using cyanamide as condensing agent, while better yields are obtained by heating certain acidic nucleotides (but so far only under conditions which are hardly likely to have occurred on the primitive earth). At the present time, more and more attention is being directed to the catalytic activity of minerals. The recent polymerizations of amino acids on Montmorillonite open up new and exciting possibilities for prebiotic synthesis.

We are convinced that natural selection, acting on a system of polymers (some of which are able to replicate), was responsible for the emergence of organized biological structures. Unfortunately, we know very little about the chemical processes involved. The operation of the Watson-Crick pairing rules in the condensation of nucleotides on preformed templates has been demonstrated, but we are far from achieving replication of an arbitrary polynucleotide. The origin of the specific amino-acid-polynucleotide interactions involved in protein synthesis is at the moment a subject for speculation.

The origin of optical specificity is a special problem in biological organization. There are good reasons why nucleic acids must contain D- or L-nucleosides, but not both; the optical purity of the amino acids in proteins is also understandable. We believe that the presence of L-amino acids and D-nucleosides in all living organisms rather than D-amino acids and L-nucleosides is an accident.

Since 1969 two types of evidence have become available showing that organic compounds are present in large quantities elsewhere in the universe. Water, ammonia, formaldehyde, hydrogen cyanide, and cyanoacetylene have all been identified in dust clouds outside the solar system. The Murchison meteorite has been shown to contain large amounts of indigenous amino acids including glycine, alanine, proline, valine, and glutamic acid. These results suggest that our notions about prebiotic synthesis have progressed along the correct lines.

It is believed that studies on prebiotic chemistry are relevant to considerations of the existence of life elsewhere in the universe. All of the available astronomical data are consistent with the possibility that there are many planets in our own and other galaxies that are suitable for the development of life. However, present astronomical techniques are not yet adequate to detect the presence of planets around a star. We suspect that life exists on some of them and that ours is not the only intelligent civilization. The initiation of a modest search program aimed at contacting other intelligent societies, if they exist, would be justified on the basis of our present understanding of the factors involved.

The reader has probably been dissatisfied with some of the chapters in this book, particularly those that are long, detailed, and inconclusive. They are this way because one or more simple processes have not yet been found. When the correct tricks are devised, the discussion will become much shorter and simplified. We expect that it will eventually be possible to demonstrate all of the required prebiotic reactions in a convincing way. We hope that this book will tempt some readers to join in the search.

Index